Hugh Miller, Harriet Myrtle

Sketch-Book of Popular Geology

Hugh Miller, Harriet Myrtle

Sketch-Book of Popular Geology

ISBN/EAN: 9783744675314

Printed in Europe, USA, Canada, Australia, Japan

Cover: Foto ©berggeist007 / pixelio.de

More available books at **www.hansebooks.com**

SKETCH-BOOK

OF

POPULAR GEOLOGY.

SKETCH-BOOK

OF

POPULAR GEOLOGY

BY HUGH MILLER,
AUTHOR OF 'THE OLD RED SANDSTONE,' ETC. ETC.

THIRD EDITION.

EDINBURGH:
WILLIAM P. NIMMO.
1869.

EDINBURGH: T. CONSTABLE,
PRINTER TO THE QUEEN, AND TO THE UNIVERSITY.

TO THE REV. W. S. SYMONDS,

RECTOR OF PENDOCK, HEREFORDSHIRE.

Dear Sir,
          Am I presuming too much on my position, as merely the editor of the following Lectures, when I ask leave to dedicate them to you? It is unquestionably a liberty with the production of another which only very peculiar circumstances can at all excuse. Yet, in the present case, I venture to think that those peculiar circumstances do exist; and I feel assured he would readily pardon me, whose work this is, and whose memory you so much revere. Without your co-operation, I believe that neither the *Cruise of the Betsey* nor these pages could by this time have seen the light. When my own over-laden brain refused to do its duty, you gave me to hope, by offers of well-timed assistance, that the task before me might still be accomplished. Your friendly voice, often heard in tones of sympathizing inquiry when I was unable to endure your own or any other human presence,—even that of my dear child,—was for a time the only sound that brought to my heart any promise or cheer for the future. It was then, while unable to read the very characters in which they were written, that I put into your hands the papers containing *The Cruise* and *Ten Thousand Miles over the Fossiliferous Deposits of Scotland.* You undertook the editorial duties connected with them *con amore*, and performed your task in a manner that left nothing to be desired.

During the preparation of the present volume for the press, you have given me all the advantage of your ready stores of information, both in carefully scrutinizing the text to see where any addition was required in the form of notes, and in referring me to the best authorities on every point regarding which I consulted you. And while so doing, you have confirmed my own judgment,—perhaps too liable to be swayed by partiality,—by expressing your conviction that this work is calculated to advance the reputation of its author.

Long may you be spared to be, as now, the life and soul of those scientific pursuits so successfully carried on in your own district! Many a happy field-day may you enjoy in connexion with that Society of which you are the honoured president. Would that all associations throughout our country were as harmless in their methods of finding recreation, as invigorating to body and mind, and as beneficial in their results to the cause of science! In exploring the beautiful fields, and woods, and sunny slopes of Worcestershire and Herefordshire, in earnest and healthful communings with nature, and, I trust, with nature's God,—the perennial springs of whose bounty are seldom quaffed in this manner as they ought to be,—I trust that much, much happiness is in store for you and for the other gentlemen of the Malvern Club,[1] to whom, as well as to yourself, I owe a debt of grateful remembrance.

And for the still higher and nobler work which God has given you to do, may He grant you no stinted measure of His abundant grace, to enable you to perform it aright.

Ever believe me, dear Sir,

Yours most faithfully,

LYDIA MILLER.

---

[1] The Malvern Club devotes stated periods,—monthly, I think,—to rambles over twenty or thirty miles of country, when the naturalists of whom it is composed,—botanists, geologists, etc.,—carry on the researches of their various departments separately, or in little groups of two or three, as they may desire. They all dine afterwards together at an inn or farmhouse, as the case may be, where they relate the adventures of the day, discuss their favourite topics, and compare their newly-found treasures. As a consequence of this, the Malvern Museum is a perfect model of what a local museum ought to be. There is no town or district of country where a few young men, possessing the advantage of an occasional holiday, might not thus associate themselves with the utmost advantage both to themselves and others.

CONTENTS.

LECTURE FIRST.

PAGE

Junction of Geologic and Human History—Scottish History of Modern Date —The Two Periods previous to the Roman Invasion: the Stone Age and the Bronze Age—Geological Deposits of these Prehistoric Periods— The Aboriginal Woods of Scotland—Scotch Mosses consequences of the Roman Invasion—How formed—Deposits, Natural and Artificial, found under them—The Sand Dunes of Scotland—Human Remains and Works of Art found in them—An Old Church disinterred in 1835 on the Coast of Cornwall—Controversy regarding it—Ancient Scotch Barony underlying the Sand—The Old and New Coast Lines in Scotland—Where chiefly to be observed—Geology the Science of Landscape—Scenery of the Old and New Coast Lines—Date of the Change of Level from the Old to the New Coast Line uncertain—Beyond the Historic, but within the Human Period—Evidences of the fact in remains of Primitive Weapons and Ancient Boats—Changes of Level not rare events to the Geologist—Some of these enumerated—The Boulder-Clay—Its prevalence in the Lowlands of Scotland—Indicated in the Scenery of the Country—The Scratchings on the Boulders accounted for—Produced by the Grating of Icebergs when Scotland was submerged—Direction in which Icebergs floated, from West to East—'Crag and Tail,' the effect of it—Probable Cause of the Westerly Direction of the Current, . . . 1-43

LECTURE SECOND.

Problem first propounded to the Author in a Quarry—The Quarry's Two Deposits, Old Red Sandstone and Boulder Clay—The Boulder-Clay formed while the Land was subsiding—The Groovings and Polishings of the Rocks in the Lower Parts of the Country evidences of the fact—Sir Charles Lyell's Observations on the Canadian Lake District—Close of the Boulder-Clay Record in Scotland—Its Continuance in England into the Pliocene Ages—The Trees and Animals of the Pre-Glacial Periods— Elephants' Tusks found in Scotland and England regarded as the Remains of Giants—Legends concerning them—Marine Deposits beneath the Pre-Glacial Forests of England—Objections of Theologians to the Geological Theory of the Antiquity of the Earth and of the Human Race considered—Extent of the Glacial Period in Scotland—Evidences of Glacial Action in Glencoe, Gareloch, and the Highlands of Sutherland—Scenery of Scotland owes its Characteristics to Glacial Action—The Period of Elevation which succeeded the Period of Subsidence—Its Indications in Raised Beaches and Subsoils—How the Subsoils and Brick Clays were formed—Their Economic Importance—Boulder-Stones interesting Features in the Landscape—Their prevalence in Scotland—The more remarkable Ice-travelled Boulders described—Anecdotes of the 'Travelled Stone of Petty' and the Standing-Stone of Torboll—Elevation of the Land during the Post-Tertiary Period which succeeded the Period of the Boulder-Clay—The Alpine Plants of Scotland the Vegetable Aborigines of the Country—Panoramic View of the Pleistocene and Post-Tertiary Periods—Modern Science not adverse to the Development of the Imaginative Faculty, ... 44-80

LECTURE THIRD.

The Poet Delta (Dr. Moir)—His Definition of Poetry—His Death—His Burial-place at Inveresk—Vision, Geological and Historical, of the Surrounding Country—What it is that imparts to Nature its Poetry—The Tertiary Formation in Scotland—In Geologic History all Ages contemporary—Amber the Resin of the *Pinus succinifer*—A Vegetable Production of the Middle Tertiary Ages—Its Properties and Uses—The Masses of Insects enclosed in it—The Structural Geology of Scotland—Its Trap Rock—The Scenery usually associated with the Trap Rock—How formed—The Cretaceous Period in Scotland—Its Productions—The Chalk Deposits—Death of Species dependent on Laws different from those which determine the Death of Individuals—The Two great Infinites, . . 81-120

LECTURE FOURTH.

The Continuity of Existences twice broken in Geological History—The three great Geological Divisions representative of three independent Orders of Existences—Origin of the Wealden in England—Its great Depth and high Antiquity—The question whether the Weald Formation belongs to the Cretaceous or the Oolitic System determined in favour of the latter by its Position in Scotland—Its Organisms, consisting of both Salt and Fresh-Water Animals, indicative of its Fluviatile Origin, but in proximity to the Ocean—The Outliers of the Weald in Morayshire—Their Organisms—The *Sabbath-Stone* of the Northumberland Coal Pits—Origin of its Name—The Framework of Scotland—The Conditions under which it may have been formed—The Lias and the Oolite produced by the last great Upheaval of its Northern Mountains—The Line of Elevation of the Lowland Counties—Localities of the Oolitic Deposits of Scotland—Its Flora and Fauna—History of one of its Pine Trees—Its Animal Organisms—A Walk into the Wilds of the Oolite Hills of Sutherland, . . . 121-152

LECTURE FIFTH.

The Lias of the Hill of Eathie—The Beauty of its Shores—Its Deposits, how formed—Their Animal Organisms indicative of successive Platforms of Existences—Laws of Generation and of Death—The Triassic System—Its Economic and Geographic Importance—Animal Footprints, but no Fossil Organisms, found in it—The Science of *Ichnology* originated in this fact—Illustrated by the appearance of the Compensation Pond, near Edinburgh, in 1842—The Phenomena indicated by the Footprints in the Triassic System—The Triassic and Permian Systems once regarded as one, under the name of the New Red Sandstone—The Coal Measures in Scotland next in order of Succession to the Triassic System—Differences in the Organisms of the two Systems—Extent of the Coal Measures of Scotland—Their Scenic Peculiarities—Ancient Flora of the Carboniferous Period—Its Fauna—Its Reptiles and Reptile Fishes—The other Organisms of the Period—Great Depth of the System—The Processes by which during countless Ages it had been formed, ... 153-194

LECTURE SIXTH.

Remote Antiquity of the Old Red Sandstone—Suggestive of the vast Tracts of Time with which the Geologist has to deal—Its great Depth and Extent in Scotland and England—Peculiarity of its Scenery—Reflection on first discovering the Outline of a Fragment of the Asterolepis traced on one of its Rocks—Consists of Three Distinct Formations—Their Vegetable Organisms—The Caithness Flagstones: how formed—The Fauna of the Old Red Sandstone—The Pterichthys of the Upper or Newest Formation—The Cephalaspis of the Lower Formation—The Middle Formation the most abundant in Organic Remains—Destruction of Animal Life in the Formation sudden and violent—The Asterolepis and Coccosteus—The Silurian the Oldest of the Geologic Systems—That in which Animal and Vegetable Life had their earliest beginnings—The Theologians and Geologists on the Antiquity of the Globe—Extent of the Silurian System in Scotland—The Classic Scenery of the Country situated on it—Comparatively Poor in Animal and Vegetable Organisms—The Unfossiliferous Primary Rocks of Scotland—Its Highland Scenery formed of them—Description of Glencoe—Other Highland Scenery glanced at—Probable Depth of the Primary Stratified Rocks of Scotland—How deposited—Speculations of Philosophers regarding the Processes to which the Earth owes its present Form—The Author's Views on the subject, . . 195-240

CONTENTS OF APPENDIX.

	PAGE
ACCUMULATIONS OF SHELLS, PHENOMENA EXPLANATORY OF,	279
AMMONITES OF THE NORTHERN LIAS,	287
ASTREA OF THE OOLITE, SUTHERLAND,	250
BELEMNITES OF THE NORTHERN LIAS,	287
BONE-BED, RECENT, IN THE FORMING,	244
BRAAMBURY, QUARRY OF, UPPER OOLITE, SUTHERLAND,	257
BREWSTER, SIR DAVID, ON THE CUTTLE-FISH AND BELEMNITE,	309
BRORA COAL-FIELD OTHER THAN THE TRUE COAL-MEASURES,	252
BRORA PEAT-MOSSES OF THE OOLITE,	256
CAUTION TO GEOLOGISTS ON THE FINDING OF REMAINS,	281
CLAY-BED OF THE NORTHERN SUTOR, LESSON TO YOUNG GEOLOGISTS,	276
CONGENERS OF THE CUTTLE-FISH, BELEMNITE, ETC.,	295
COPROLITES OF THE LIAS,	303
CROMARTY,	268
CROMARTY, CAVES OF, OR THE ART OF SEEING OVER THE ART OF THEORIZING,	269
CROMARTY SUTOR, LINE OF,	275
CUTTLE-FISH,	288
DIPTERUS MACROLEPIDOTUS, ABUNDANT IN THE BANNISKIRK OLD RED OF CAITHNESS,	246
EATHIE, INTRUSIVE DIKES OF,	304

CONTENTS.

	PAGE
ECONOMIC GEOLOGY, LONDON MUSEUM OF,	254
FOSSIL-WOOD OF THE OOLITE AT HELMSDALE, SUTHERLAND,	247
GANOID SCALES AND RAYS,	243
GLACIAL APPEARANCES AT NIGG AND LOGIE,	277
GLACIERS AND MORAINES OF SUTHERLAND,	260
GRANITIC GNEISS AND SANDSTONE, WITH THE CONDITIONS OF THEIR UPHEAVAL,	284
LEVEL STEPPES OF RUSSIA, AND THEORY OF MORAINES,	265
SEPTARIA, OR CEMENT-STONES, OF THE LIAS,	286
TEREBRATULA, CONTEMPORARY AND EXTINCT TYPES OF THE LIFE OF,	307
TRAVELLED BOULDERS NOT ASSOCIATED WITH CLAY,	283
TYPES, RECENT, OF FOSSILS,	251
UNDERLYING CLAY ON LEVEL MOORS, REMARKS ON,	282

THEORY OF THE OCEAN'S LEVEL,	312
CHAIN OF CAUSES,	320
RECENT GEOLOGICAL DISCOVERIES,	344
SIR RODERICK MURCHISON ON THE RECENT GEOLOGICAL DISCOVERIES IN MORAYSHIRE,	353

PREFACE.

THE following Lectures, with *The Cruise of the Betsey*, and *Rambles of a Geologist*, are all that remain of what Hugh Miller once designed to be his *Maximum Opus*,— THE GEOLOGY OF SCOTLAND. It is well, however, that his materials have been so left that they can be presented to the public in a shape perfectly readable; furnishing two volumes, each of which, it is hoped, will be found to possess in itself a uniform and intrinsic interest,—differing in matter and manner as much as they do in the form in which they have found an embodiment. That form is simply the one naturally arising out of the circumstances of the Author's life as they occurred, instead of the more artificial plan designed by himself, in which these circumstances would probably more or less, if not altogether, have disappeared. Yet it may well be doubted whether the natural method does not possess a charm which any more formal arrangement would have wanted. Every one must be struck with the freshness, buoyancy, and vigour displayed in the *Summer Rambles*; qualities more apparent in these than even in his more laboured Autobiography, of which they are, indeed, but a sort of unintentional continuation. They were the spontaneous utterances of a mind set free from an occupation never very congenial,—that of writing compulsory articles for a newspaper,—to find refreshment amid the familiar haunts in which it delighted, and to seize with a grasp, easy, yet powerful, on the recreation of a favourite science, as the artist seizes on the pencil from which he has been separated for a time, or the musician on some instrument much loved and long lost, which he well knows will, as it yields to him its old music, restore vigour and harmony to his entire being. My dear husband did, indeed, bring to his science all that fondness, while he found in it much of that kind of enjoyment, which we are wont to associate exclusively with the love of art.

The delivery of these Lectures may not yet have passed quite away from the recollection of the Edinburgh public. They excited unusual interest, and awakened unusual attention, in a city where interest in scientific matters, and attendance upon lectures of a very superior order, are affairs of every-day occurrence. Rarely have I seen an audience so profoundly absorbed. And at the conclusion of the whole, when the lecturer's success had been triumphantly established (for it must be remembered that lecturing was to him an *experiment* made late in life), I ventured to urge the propriety of having the series published before the general interest had begun to subside. His reply was, 'I cannot afford it : I have given so many of my best facts and broadest ideas,—so much, indeed, of what would be required to lighten the drier details in my *Geology of Scotland*,—that it would never do to publish these Lectures by themselves.' It will thus be seen that they veritably gather into one luminous centre the best portions of his contemplated work, garnering very much of what was most vivid in painting and original in conception,—of that which has now, alas ! glided, with himself, into those silent shades where dwell the souls of the departed, with the halo of past thought hovering dimly round them, waiting for that new impulse from the Divine Spirit which is to quicken them into an intenser and higher unity.

I have been led to indulge the hope that this work will be found useful in giving to *elementary* Geology a greater attractiveness in the eyes of the student than it has hitherto possessed. It was characteristic of the mind of its author that he valued words, and even facts, as only subservient to the high powers of reason and imagination. It is to be regretted that many introductory works, especially those for the use of schools, should be so crammed with scientific terms, and facts hard packed, and not always well chosen, that they are fitted to remind us of the dragon's teeth sown by Jason, which sprang up into armed men,—being much more likely to repel, than to allure into the temple of science. One might, indeed, as well attempt to gain an acquaintance with English literature solely from the study of Johnson's Dictionary, as to acquire an insight into the nature of Geology from puzzling over such books. But, viewed in the light of a mind which had approached the subject by quite another pathway, all unconscious, in its outset, of the gatherings and recordings of others, and which never made a single step of progression in which it was not guided by the light of its own genius and the inspiration of nature, it may be regarded by beginners in another aspect,—one very different from that in which Wordsworth looked upon it when he thanked Heaven that the covert nooks of nature reported not of the geologist's hands,—'the man who classed his splinter by some barbarous name, and hurried on.' At that time the poet must have seen but the cold, hard profile of the man, instead of the broad, beaming, full-orbed glance which he may cast over the wondrous æons of the past eternity.

To meet any difficulties arising from misconception, it may be proper to glance rapidly at what has been accomplished in geological research within the last two years. The reader will thus avoid the painful impression that there are any suppressed facts of recent date which clash with the theories of the succeeding Lectures, destroying their value and impairing their unity. And it may be well to remind him that there are two schools of Geology, quite at one in their willingness to bring all theories to the test of actual discovery, but widely differing in their leanings as to the mode in which, *a priori*, they would wish the facts brought to light to be viewed. The one, as expounded in the following Lectures, delights in the unfolding of a great plan, having its original in the Divine Mind, which has gradually fitted the earth to be the habitation of intelligent beings, and has introduced upon the stage of time organism after organism, rising in dignity, until all have found their completion in the human nature, which, in its turn, is a prophecy of the spiritual and Divine. This may be said to be the *true* development hypothesis, in opposition to the false and puerile one, which has been discarded by all geologists worthy of the name, of whatsoever side. The other school holds the opinion,—though perhaps not very decidedly,—that all things have been from the beginning as they are now; and that if evidence at the present moment leans to the side of a gradual progress and a serial development, it is because so much remains undiscovered; the hiatus, wherever it occurs, being always in our own knowledge, and not in the actual state of things. The next score of years will probably bring the matter to a pretty fair decision; for it seems impossible that, if so many able workers continue to be employed as industriously as now in the same field, the remains of man and the higher mammals will not be found to be of all periods, if at all periods they existed. In the meantime, it is well to know the actual point to which discovery has conducted us; and this I have taken every pains most carefully to ascertain.

The Upper Ludlow rocks,—the uppermost of the Silurians,—continue to be the lowest point at which fish are found. Up to that period,—during the vast ages of the Cambrian, where only the faintest traces of animal life have been detected[1] in the shape of annelides or sand-boring worms,—throughout the whole range of the Silurians, where shell-fish and crustaceans, with inferior forms of life, abounded,—no traces of fish, the lowest vertebrate existences until the latest formed beds of the Upper Silurian, have yet appeared. There are now six genera of fish ranked as Upper Silurian,—Auchenaspis, Cephalaspis, Pteraspis, Plectrodus, Onchus Murchisoni, and Sphagodus. The two latter,—Onchus Murchisoni and Sphagodus,—are represented by bony defences, such as are possessed by placoid fishes of the present day. Sir Roderick Murchison at one time entertained the idea of placing the Ludlow bone-bed at the base of the Old Red Sandstone; but its fish having been found decidedly associated with Silurian organisms, this idea has been abandoned.

[1] See the lately published edition of Sir Roderick I. Murchison's *Siluria*, chap. ii. p. 26.

The next point to which public attention has been specially directed is the discovery of mammals lower than they had formerly appeared. Considerable misconception has arisen on this head. The Middle Purbeck beds, recently explored by Mr. Beckles, in which various small mammals were found, occur considerably farther up than the Stonesfield slates, in which the first quadruped was detected so far back as 1818. But this discovery involves no theoretical change, inasmuch as all the mammalian remains of the Middle Purbecks consist of small marsupials and insectivora, varying in size from a rat to a hedgehog, with one or two doubtful species, not yet proved to be otherwise. The living analogue of one very interesting genus is the kangaroo rat, which inhabits the prairies and scrub-jungles of Australia, feeding on plants and scratched-up roots. Between the English Stonesfield or Great Oolite, in which many years ago *four* species of these small mammals were known to exist, and the Middle Purbeck, quarried by Mr. Beckles, in which *fourteen* species are now found, there intervene the Oxford Clay, Coral Rag, Kimmeridge Clay, Portland Oolite, and Lower Purbeck Oolite; and then, after the Middle Purbeck, there occurs a great hiatus throughout the Weald, Green Sand, Gault, and Chalk, wherein no quadrupedal remains have been found; until at length we are introduced, in the Tertiary, to the dawn of the grand mammalian period; so that nothing has occurred in this department to occasion any revolution in the ideas of those who, with my husband, consider a succession and development of type to be the one great fixed law of geological science. The reader will see that in the end of Lecture Third such remains as have been found lower than the Tertiary are expressly recognised and excepted. 'Save,' says the author, 'in the dwarf and inferior forms of the marsupials and insectivora, not any of the honest mammals have yet appeared.'

But while attaching no importance to the discoveries in the Middle Purbeck, except in regard of more ample numerical development, it is necessary to admit the evidence of marsupials having been found lower than the Stonesfield or Great Oolite: even so far back as the Upper Trias, the Keuper Sandstone of Germany, which lies at the base of the Lias. I must be permitted, on this point, to quote the authority of Sir Roderick Murchison, as one of the safest and most cautious exponents of geological fact. 'In that deposit,' says he, referring to the Keuper Sandstone of Würtemberg, 'the relics of a solitary small marsupial mammal have been exhumed, which its discoverer, Plieninger, has named *Microlestes Antiquus*. Again, Dr. Ebenezer Emmons, the well-known geologist of Albany, in the United States, has described, from the lower beds of the Chatham Secondary Coal-field, North Carolina (of the same age as those of Virginia, and probably of the Würtemberg Keuper), the jaws of another minute mammal, which he calls *Dromotherium Sylvestre*. Lastly, *while I write*, Mr. C. Moore has detected in an agglomerate which fills the fissures of the carboniferous limestone near Frome, Somersetshire, the teeth of marsupial mammals, one of which he considers to be closely related to the *Microlestes Antiquus* of Germany, and Professor Owen confirms the fact. From that coincidence, and also from the association with other animal remains,—the Placodus (a reptile of the Muschelkalk), and certain mollusca,—Mr. Moore believes that these patches represent the Keuper of Germany. If this view should be sustained, this author, who has already made remarkable additions to our acquaintance with the organic remains of the Oolitic rocks and the Lias, will have had the merit of having discovered the first traces of mammalia in any British stratum below the Stonesfield slates.' . . . 'Let me entreat,' says Sir Roderick, in a passage occurring shortly after that we have quoted,—'Let me entreat the reader not to be led, by the reasoning of the ablest physiologist, or by an appeal to minute structural affinities, to impugn the clear and exact facts of a succession from lower to higher grades of life in each formation. Let no one imagine that because the bony characters in the jaw and teeth of the Plagiaulax of the Purbeck strata are such as the comparative anatomist might have expected to find among existing marsupials, and that the animal is therefore far removed from the embryonic archetype, such an argument disturbs the order of succession of *classes*, as seen in the crust of the earth.' So far from disturbing the order of succession, it is, we conceive, of exceeding interest to find the Mesozoic period marked in its commencement, as it most probably will be found to be, by the introduction of a form of being so entirely different from any that preceded it. It seems to us to bring the true development hypothesis into a clearer and more harmonious unity. The great period during which the little annelide or sand-boring worm was the sole tenant of this wide earth,—its first inhabitant after the primeval void,—has passed. The æon of the Mollusc and the Crustacean follows. At its close appear the first fishes, very scanty in point of numbers and of species, but multiplying into many genera, and swarming in countless myriads, as the Devonian ages wear on. Again, towards the termination of the latter appear the first reptiles, which, during the Carboniferous and Permian eras, reign as the master-existences of creation. But Palæozoic or ancient life passes away, and the Mesozoic or Middle period is marked, not only by countless forms, all specifically, and many of them generically, new, but by another wholly unknown, either as genus or species, during all the past. The little marsupials and insectivora appear 'perfect, after their kind,' and yet only the harbingers of the great mammalian period which is yet to come. In the volume of Creation, as in that of Providence, God's designs are wrapt in profound mystery until their completion. And yet in each it would appear that He sends a prophetic messenger to prepare the way, in which the clear-sighted eye, intent to read His purposes, may discern some sign of the approaching future.

Before we proceed, we must here, on behalf of the unlearned, and therefore the more easily misled, most humbly venture to reclaim against the use, on the part of men of the very highest standing, of the loose and dubious phraseology in which they sometimes indulge, and which serves greatly to perplex, if not to lead to very erroneous conclusions.

'In respect to no *one class* of animals,' says Professor Owen, in his last Address to the British Association, 'has the manifestation of creative force been limited to one epoch of time.' This, translated into fact, can only mean that the vertebrate type had its representative in the fish of the earliest or Silurian epoch, and has continued to exist throughout all the epochs which succeeded it. But the difficulty lies in the translation. For at first sight the conclusion is inevitable to the general reader, that not only the lowest class of vertebrate existence, but also man and the higher mammals, had been found from the beginning, and that the highest and the lowest forms of being were at all periods contemporary. No one surely would have a right to make such a prodigious stride in the line of inference, on the presumption of supposed evidence yet to come. Again, Sir Charles Lyell, in his supplement to the fifth edition of his *Elementary Geology*, says, in speaking of these same Purbeck beds quarried by Mr. Beckles, 'They afford the first positive proof as yet obtained of the co-existence of a varied fauna of the *highest class* of vertebrata with that ample development of reptile life which marks the periods from the Trias to the Lower Cretaceous inclusive.' *Are* marsupials and insectivora the *highest class* of vertebrata? Where, then, do the great placental mammals,—where does man himself,—take rank?

It were surely to be desired that some stricter and more invariable form of phraseology were adopted, either in accordance with the divisions of Cuvier, or some analogous system, adherence to which would be clearly defined and understood. Why should not the words *class, order, type*, have as invariable a meaning as *genera* and *species*, which, having an application more limited, are seldom mistaken? We are aware that such terms are often used by the learned in an indefinite and translatable sense, just as to the learned in languages it may be a matter of indifference whether the written characters which convey information to them be Roman, Hebrew, or Chinese. But it should be remembered that there is a large class outside which seeks to be addressed in a plain vernacular,—which asks, first of all, definiteness in the use of terms to which probably they have already sought to attach some fixed sense; and that it is not well to unship the rudder of their thought, and send them back to sea again.

The next point which demands attention in our short *résumé* is that great break between the Permian and Triassic systems, across which, as stated in the following pages, not a single *species* has found its way. Much attention has been given to the great Hallstad or St. Cassian beds, which lie on the northern and southern declivities of the Austrian Alps. These beds belong to the *Upper Trias*, and they contain more *genera* common to Palæozoic and newer rocks than were formerly known. There are ten genera peculiarly Triassic, ten common to older, and ten to newer strata. Among these, the most remarkable is the Orthoceras, which was before held to be altogether Palæozoic, but is here found associated with the Ammonites and Belemnites of the secondary period.[1] The appearance of this, with a few other familiar forms, serves, in our imagination at least, to lessen the distance, and, in some small measure, to bridge over the chasm, between Palæozoic and Secondary life. And yet, considering the vast change which then passed over our planet,—that all specific forms died out, while new ones came to occupy their room,—the discovery of a few more connecting generic links in the rudimentary shell-alphabet, which serve but to show that in all changes the God of the past is likewise the God of the present, no more affects in reality this one great revolution,

---

[1] See Sir Charles Lyell's *Supplement* for corroboration of the foregoing statements.

the completeness of which is marked by the very difficulty of finding, amid so much new and redundant life, a single identical specific variety, than the well-known existence of the Terebratula in the earliest, as well as in the existing seas, can efface the great ground-plan of successive geological eras.[1] Nor does it explain the matter to say that geographical changes took place, bringing with them the denizens of different climates, and adapted for different modes of life. The same Almighty Power which *now* provides habitats and conditions suitable for the wants of his creatures, would doubtless have done so during all the past. Geographical changes are at all times indissolubly connected with changes in the conditions of being; and they serve, in so far, to explain the *rule* in the stated order of geological events, when a due proportion of extinct and of novel forms are found co-existent. But how can they explain the exception? A singular effect must have a singular cause. And when we find that there were changes relating to the world's inhabitants altogether singular and abnormal in their revolutionary character, we must infer that the medial causes of which the Creator made use were of a singular and abnormal character also. On this head the best-informed ought to speak with extreme diffidence. We can but imagine that there may have been a long, immeasurable period during which a subsidence, so to speak, took place in the creative energy, and during which all specific forms, one after another, died out,—the lull of a dying creation,—and then a renewal of the impulsive force from that Divine Spirit which brooded over the face of the earliest chaotic

---

[1] See *Terebratula*, in Appendix. The extinct Terebratula is now called Rhynconella.

deep, producing geographical changes, more or less rapid, which should prepare the way for the next stage in our planetary existence,—its new framework, and its fresh burden of vital beings.

The other great break in the continuity of fossils, which occurs between the Chalk and the Tertiary, seems to be very much in the same condition with that of which we have just spoken. New connecting *genera* have been discovered, but still not a single identical species. Jukes, in his *Manual*, published at the end of last year, says,—' Near Maestricht, in Holland, the chalk, with flint, is covered by a kind of chalky rock, with grey flints, over which are loose yellowish limestones, sometimes almost made up of fossils.' Similar beds also occur at Saxoe in Denmark. Together with true cretaceous fossils, such as pecten and quadricostatus, these beds contain species of the genera Voluta, Fasciolaria, Cyprea, Oliva, etc. etc., several of which GENERA are only found elsewhere in the Tertiary rocks.[1]

Sir Roderick Murchison's late explorations in the Highlands,—although, of course, local in their character,—have made a considerable change in the GEOLOGY OF SCOTLAND. The next edition of the *Old Red Sandstone* will be the most fitting place to speak of these at length; and I have some reason to believe that Sir Roderick himself will then favour me with a communication giving some account of them. Suffice it at present to say, that the supposed Old Red Conglomerate of the Western Highlands, as laid down in the year 1827 by Sir Roderick himself, accompanied by Professor Sedgwick, and so far acquiesced in by my husband,

[1] A doubt has nevertheless been expressed whether these are not broken-up Tertiaries.

although he always wrote doubtfully on the subject, has now been ascertained to be, not Old Red, but Silurian. In Sir Roderick's last Address to the British Association, he says,—'Professor Sedgwick and himself had thirty-one years ago ascertained an ascending order from gneiss, covered by quartz rocks, with limestone, into overlying quartzose, micaceous, and other crystalline rocks, some of which have a gneissose character. They had also observed what they supposed to be an associated formation of red grit and sandstone; but the exact relations of this to the crystalline rocks was not ascertained, owing to bad weather. In the meantime, *they, as well as all subsequent geologists, had erred in believing* the great and lofty masses of purple and red conglomerate of the western coast were of the same age as those on the east, and therefore 'Old Red Sandstone.' . . . 'Professor Nicol had suggested that the quartzites and limestones might be the equivalent of the Carboniferous system of the south of Scotland. Wholly dissenting from that hypothesis, he (Sir Roderick) had urged Mr. Peach to avail himself of his first leisure moments to re-examine the fossil-beds of Durness and Assynt, and the result was the discovery of so many forms of undoubted Lower Silurian characters (determined by Mr. Salter), that the question has been completely set at rest, there being now no less than nineteen or twenty species of M'Lurea, Murchisonia, Cephalita, and Orthoceras, with an Orthis, etc., of which ten or eleven occur in the Lower Silurian rocks of North America.'

This change would demand an entirely new map of the Geology of Scotland; for there is clearly ascertained to be an ascending series from west to east, beginning with an older or primitive gneiss, on which a Cambrian conglomerate, and over that again a band containing the Silurian fossils, rest; while a younger gneiss occupies a portion of the central nucleus, having the Old Red Sandstone series on the eastern side. A change has likewise been made in the internal arrangements of the Old Red, of which the next edition of my husband's work on the subject will be the proper place to speak in detail. In the meantime, I may just mention, that the Caithness and Cromarty beds have been found to occupy, not the lowest, but the central place, the lowest being assigned to the Forfarshire beds, containing Cephalaspis, associated with Pteraspis, an organism characteristically Silurian. That which bears most upon the subject before us is the now perfectly ascertained imprint of the footsteps of large reptiles in the Elgin or uppermost formation of the Old Red. A shade of doubt had rested upon the discovery made many years ago by Mr. Patrick Duff of the *Telerpeton Elginense*, not as to the real nature of the fossil, which is indisputably a small lizard, but as to whether the stratum in which it was found belonged to the Old Red, or to the formation immediately above it. It will be observed, however, that the existence of reptiles in the Old Red did not rest altogether upon this, because the footprints of large animals of the same class had been ascertained in the United States of America. I cannot but conceive, therefore, that Mr. Duff, in a recent letter or paper read in Elgin, and published in the *Elgin and Morayshire Courier*, makes too much of the recent discoveries in his neighbourhood, when he asserts that the Old Red Sandstone has been hitherto considered exclusively a *fish* formation, and that the appearance of reptiles is altogether novel.

'Now,' says he, 'that the Old Red Sandstones of Moray have acquired some celebrity, it may not be unprofitable to trace the different stages by which the discovery was arrived at of reptilian remains in that very ancient system, *which till now was held to have been peopled by no higher order of beings than fishes.*' Mr. Duff forgets that in the programme, as it may be called, given by my husband, of the introduction of different types of animal life, as ascertained in his day, reptiles are made to occupy precisely the position they do now. To refresh the memory of the reader, I shall here reproduce it, as given in the *Testimony of the Rocks.* At page 14 is this diagram :—

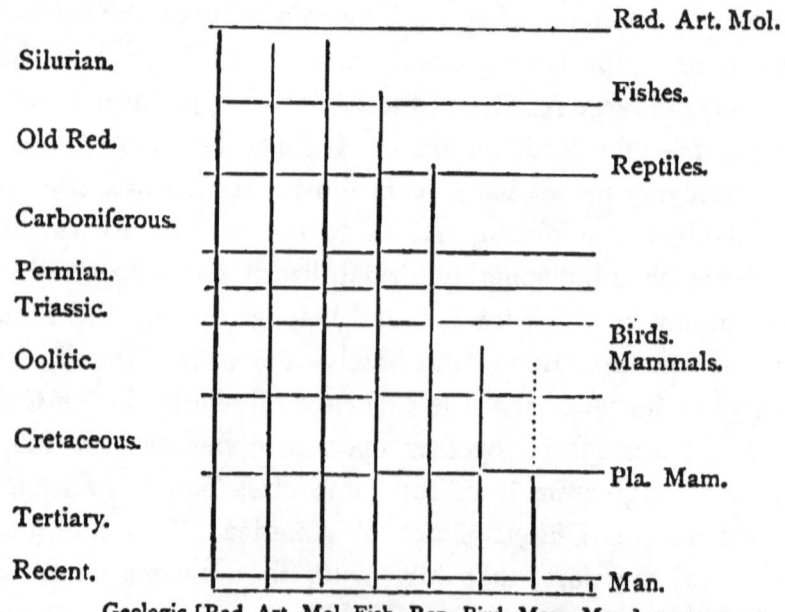

Geologic [Rad. Art. Mol. Fish. Rep. Bird. Mam. Man.] arrangement.
Cuvier's [Rad. Art. Mol. Fish. Rep. Bird. Mam. Man.] arrangement.

THE GENEALOGY OF ANIMALS.

And immediately following it occurs this comment :—' In the many-folded pages of the Old Red Sandstone, *till we reach the highest and last*, there occur the remains of no other vertebrates than those of this fourth class [fishes]; but in its uppermost deposits there appear traces of the third or reptilian class; and in passing upwards still, through the Carboniferous, Permian, and Triassic systems, we find reptiles continuing the master-existences of the time.' At pages 16, 17, express allusion is made to the *Telerpeton Elginense*, with the doubt as to the nature of its *locale* very slightly touched upon.[1] All this Mr. Duff has forgotten, apparently; and it appears likewise not to have come within his cognizance that Sir Charles Lyell distinctly recognises his Telerpeton as well as the American footprints, and assigns both their proper places, in the last edition of his *Principles*. Even in the edition *before the last* of the *Siluria*, almost the first thing that meets us, on opening it at Chapter Tenth, which treats of the Old Red Sandstone, is a print of the fossil skeleton of this same *Telerpeton Elginense*,—its true place assigned to it with quite as much certainty as now! These very singlar lapses in memory seem not to be peculiar to Mr. Duff. I have seen it stated in an anonymous article published in a widely circulated journal,[2] and in connexion with the discovery of the Elgin reptile foot-prints, that Hugh Miller considered the Old Red Sand-

[1] This doubt, I see by Sir Roderick Murchison's latest Address to the British Association, is not yet entirely obviated. See Appendix.

[2] For this article, as an excellent specimen of its class, see Appendix, under the head 'Recent Geological Discoveries;' and, in contradistinction to it, the extract from Sir R. Murchison's Address ought to be carefully studied. I myself had seen neither that extract nor the recent *Siluria* until after this short sketch was in type; the references to the latter having been introduced afterwards; and it may be conceived with what feelings of gratification I have perused Sir Roderick's repeated assurances of adherence to the 'Old Light.'

stone to have been a shoreless ocean without a tree!—utterly ignoring the fact that he was himself the discoverer of the first Old Red fossil-wood of a coniferous character, and that he thence expressly infers the then existence of vegetation of a high order. Is it not enough to add to the store of knowledge without attempting to undermine all that has gone before? Must the discovery of an additional reptile, a few additional marsupials, be the signal for the immediate outcry, 'All is changed; the former things have passed away; all things have become new'? My husband was solicitous even to the point of nervous anxiety to exclude from his writings every particle of error, whether of facts or of the conclusions to be drawn from them. Much rather would he never have written at all than feel himself in any degree a false teacher. 'Truth first, come what may afterwards,' was his invariable motto. In the same spirit, God enabling me, I have been desirous to carry on the publication of his posthumous writings. God forbid that one intrusted with such sacred guardianship should seek to pervert or suppress a single truth, actual or presumptive, even though its evidence were to overthrow in a single hour all his much-loved speculations,—all his reasonings, so long cogitated, so conscientiously wrought out. Yet I must confess that I was at first startled and alarmed by rumours of changes and discoveries which, I was told, were to overturn at once the science of Geology as hitherto received, and all the evidences which had been drawn from it in favour of revealed religion. Though well persuaded that at all times, and by the most unexpected methods, the Most High is able to assert Himself, the proneness of man to make use of every unoccupied position in order to maintain his independence of his Maker seemed about to gain new vigour by acquiring a fresh vantage-ground. The old cry of the eternity of matter, and the 'all things remain as they were from the beginning until now,' rung in my ears. *God with us*, in the world of science henceforth to be no more! The very evidences of His being seemed about to be removed into a more distant and dimmer region, and a dreary swamp of infidelity spread onwards and backwards throughout the past eternity.

Without stopping to inquire whether, although the science of Geology *had* been revolutionized, those fears were not altogether exaggerated, it is enough at present to know, that as Geology has *not* been revolutionized, there is no need to entertain the question. I trust I have at least succeeded in furnishing the reader with such references,—few and simple when we once know where to find them,—as may enable him to decide upon this important matter for himself. If I have learned anything in the course of the investigations which I have been endeavouring to make, it is to take nothing upon credence, but to wait patiently for all the evidence which can be brought to bear upon the subject before me; and this, I believe, is the only way to make any approximation to a correct opinion. In truth, the science of Geology is itself in that condition, that no fact ought to be accepted as a basis for reasoning of a solid kind, until it has run the round of investigation by the most competent authorities, and has stood the test of time. It is peculiarly subject to the cry of lo, here! and lo, there! from false and imperfectly informed teachers; and I believe the men most thoroughly to be relied on are those who are the slowest to theorize, the last to form a judgment, and who require the largest amount of evidence before that judgment is finally pronounced.

In addition to the inspection of my ever kind and generous friend Mr. Symonds,[1] I have submitted the following pages to the reading of Mr. Geikie[2] of the Geological Survey, who has here and there furnished a note. Of the amount and correctness of his knowledge, acquired chiefly in the field and in the course of his professional duties, my husband had formed the highest opinion. Indeed, I believe he looked upon him as *the* individual who would most probably be his successor as an exponent of Scottish Geology. One who walks on an average twenty miles per day, and who has submitted nearly every rood of the soil to the accurate inspection demanded by the Survey, must be one whose opinion, in all that pertains to Scottish Geology in especial, must be well worth the having. I have to add an expression of most grateful thanks to Sir Roderick Murchison, for his prompt attention to sundry applications which I was constrained to make to him. His letters have been of the utmost importance in enabling me to perceive clearly the alterations which have taken place in our Scottish Geology, and the reasons for them. One feels instantaneously the benefit of contact with a master-mind. A few sentences, a few strokes of the pen, throw more light on the subject than volumes from an inferior hand.

It remains now only to explain that this course of Lectures, as delivered before the Philosophical Institution, consisted

---

[1] The Rev. W. S. Symonds, author of *Old Stones, Stones of the Valley*, etc., and the compiler of the index to the recent edition of Sir R. Murchison's *Siluria*.

[2] Archibald Geikie, Esq., author of *The Story of a Boulder*.

of eight, instead of six. Those now published are complete, according to their limits, in all that relates to the facts, literal or picturesque, of the subject; and the last two of the series will be found in *The Testimony of the Rocks*, under the heads of 'Geology in its Bearing on the Two Theologies,' and 'The Mosaic Vision of Creation.' If it had been within the contemplation of the author to publish the six Lectures as they now stand, these last two would have formed their natural climax or peroration. And, accordingly, I entertained some thought of republishing them here, in order that the reader might enjoy the advantage of having the whole under his eye at once. But as they are not in any way necessary to the completion of the sense, and perhaps Geology, viewed simply by itself, and in the light of a popular study, is as well freed from extraneous matter, it was thought best, on the whole, to refer the reader who wishes to see the eight discourses in their original connexion, to *The Testimony of the Rocks*.

I have, instead, added an Appendix of rather a novel character. In addition to the *Cruise of the Betsey*, and *Ten Thousand Miles over the Fossiliferous Deposits of Scotland*, there was left a volume of papers unpublished as a whole, entitled *A Tour through the Northern Counties of Scotland*. They had, however, been largely drawn upon in various other works; but, scattered throughout, were passages of more or less value which I had not met with elsewhere; and some such, of the descriptive kind, I have culled and arranged as an Appendix; first, because I was loath that any original observation from that mind which should never think again for the instruction of others should be lost, and also because many of those passages were of a kind which might prove suggestive to the student, and assist him in reasoning upon those phenomena of ordinary occurrence, without close observation of which no one can ever arrive at a successful interpretation of nature. If the reader should descry aught of repetition which has escaped my notice, I must crave his indulgence, in consideration of the very difficult and arduous task which God, in His mysterious providence, has allotted me. To endeavour to do by these writings as my husband himself would if he were yet with us, —to preserve the integrity of the text, and in dealing with what is new, to bring to bear upon it the same unswerving rectitude of purpose in valuing and accepting every iota of truth, whether it can be explained or not, rejecting all that is crude, and abhorring all that is false,—this has been my aim, although, alas! too conscious throughout of the comparative feebleness of the powers brought to bear upon it. If, however, the reader is led to inquire for himself, I trust he will find that these powers, such as they are, have been used in no light or frivolous spirit, but with a deep and somewhat of an adequate, sense of the vast importance of the subject.

L. M.

LECTURES ON GEOLOGY.

LECTURES ON GEOLOGY.

LECTURE FIRST.

Junction of Geologic and Human History—Scottish History of Modern Date—The Two Periods previous to the Roman Invasion: the Stone Age and the Bronze Age—Geological Deposits of these Prehistoric Periods—The Aboriginal Woods of Scotland—Scotch Mosses consequences of the Roman Invasion—How formed —Deposits, Natural and Artificial, found under them—The Sand Dunes of Scotland—Human Remains and Works of Art found in them—An Old Church disinterred in 1835 on the Coast of Cornwall—Controversy regarding it— Ancient Scotch Barony underlying the Sand—The Old and New Coast Lines in Scotland—Where chiefly to be observed—Geology the Science of Landscape —Scenery of the Old and New Coast Lines—Date of the Change of Level from the Old to the New Coast Line uncertain—Beyond the Historic but within the Human Period—Evidences of the fact in remains of Primitive Weapons and Ancient Boats—Changes of Level not rare events to the Geologist—Some of these enumerated—The Boulder-Clay—Its Prevalence in the Lowlands of Scotland—Indicated in the Scenery of the Country—The Scratchings on the Boulders accounted for—Produced by the grating of Icebergs when Scotland was submerged—Direction in which Icebergs floated, from West to East— 'Crag and Tail:' the effect of it—Probable Cause of the Westerly Direction of the Current.

IN most of the countries of Western Europe, Scotland among the rest, geological history may be regarded as ending where human history begins. The most ancient portions of the one piece on to the most modern portions of the other. But their line of junction is, if I may so express myself, not an abrupt, but a shaded line; so that, on the one hand, the human period passes so entirely into the geological, that we found our conclusions respecting the first human inhabitants rather on what may be deemed *geologic* than on the ordinary *historic* data; and, on the other hand, some of the later and lesser geologic changes have taken place in periods comparatively so recent, that, in even our own country, we are able to catch a glimpse of them in the first dawn of history proper,—that written history in which man records the deeds of his fellows.

In Scotland the ordinary historic materials are of no very ancient date. Tytler's History opens with the accession of Alexander III. in the middle of the thirteenth century; the Annals of Lord Hailes commence nearly two centuries earlier, with the accession of Malcolm Canmore; there still exist among the muniments of Durham Cathedral charters of the 'gracious Duncan,' written about the year 1035; and it is held by Runic scholars that the Anglo-Saxon inscription on the Ruthwell Cross may be about two centuries earlier still. But from beyond this comparatively modern period in Scotland no written document has descended, or no *native* inscription decipherable by the antiquary. A few votive tablets and altars, lettered by the legionaries of Agricola or Lollius Urbicus, when engaged in laying down their long lines of wall, or rearing their watch-towers, represent a still remoter period; and a few graphic passages in the classic pages of Tacitus throw a partial and fitful light on the forms and characters of the warlike people against which the ramparts were cast up, and for a time defended. But beyond this epoch, to at least the historian of the merely literary type, or to the antiquary of the purely documentary one, all is darkness. 'At one stride comes the dark.' The period is at once reached which we find so happily described by Coleridge. 'Antecedently to all history,' says the poet, 'and long glimmering through it as a hazy tradition, there presents itself to our imagination an indefinite period, dateless as eternity,—a *state* rather than a *time*. For even the sense of succession is lost in the uniformity of the stream.'

It is, however, more than probable that the age of Agricola holds but a midway place between the present time and the time in which Scotland first became a scene of human habitation. Two great periods had passed ere the period of the Roman invasion,—that earliest period now known to the antiquary as the '*stone age*,' in which the metals were unknown, and to which the flint arrow-head and the greenstone battle-axe belong; and that after-period known to the antiquary as the '*bronze age*,' in which weapons of war and the chase were formed of a mixture of copper and tin. Bronze had in the era of Agricola been supplanted among the old Caledonians by iron, as stone had at an earlier era been supplanted by bronze; and his legionaries were met in fight by men armed, much after the manner of their descendants at Sheriffmuir and Culloden, with broadsword and target. And it is known that nearly a century and a half earlier, when Cæsar first crossed the Channel, the Britons used a money made of iron. The two earlier periods of bronze and stone had come to a close in the island ere the commencement of the Christian era; and our evidence regarding them is, as I have said, properly of a geologic character. We read their history in what may be termed the *fossils* of the antiquary. Man is peculiarly a tool-and-weapon-making animal; and his tools and weapons represent always the stage of civilisation at which he has arrived. First, stone is the material out of which he fashions his implements. If we except that family of man which preserved the aboriginal civilisation, there seems never to have been a tribe or nation that had not at one time recourse to this most obvious of substances for their tools and weapons. Then comes an age in which stone is supplanted by the metals that occur in a native state,—*i.e.*, in a state of ductility in the rock,—such as copper, silver, and gold. Of these, copper is by much the most abundant; and in all countries in which it has been employed for tools and weapons means have been found by the primitive workers to harden it through an admixture of other metals, such as zinc and tin. Last of all, the comparatively occult art of smelting iron is discovered, and the further art of converting it into steel; and such is its superiority in this form to every other metal employed in the fabrication of implements, that it supplants every other; and the battle-axe and chisel of hardened copper (bronze) are as certainly superseded by it as the chisel and the battle-axe of stone had at an earlier period been superseded by the bronze.[1] Now, it is truly wonderful how thoroughly, for all general purposes, this scheme of classification, which we owe to the Danish antiquary Thomsen, arranges into corresponding sections and groups the antiquities of a country, and gives to it a legible history in ages unrecorded by the chronicler. With the stone tools or weapons there are found associated in our own country, for instance, a certain style of sepulture, a certain type of cranium, a certain form of human dwellings, a certain class of personal ornaments, certain rude log-hollowed canoes, undressed standing stones, and curiously-poised cromlechs. The bronze tool or weapon has also its associated class of antiquities,—massive ornaments of gold,

[1] In an interesting article on Ireland which lately appeared in the *Scotsman* newspaper, I find it stated that for a very considerable distance, ' between Lough Rea and Lough Derg, the river Shannon was fordable at only one point, which of course formed the only medium of communication between the natives of the two banks. They seem, however,' it is added, 'to have met oftener for war than peace; and from this ford a whole series of ancient warlike weapons was dug out. These weapons are now preserved in the fine collection of antiquities in the Museum of the Royal Irish Academy in Dublin, and are partly bronze and partly stone. Their position in the river bed told a curious tale, both historically and geologically. The weapons of bronze were all found in the upper stratum, and below them those of stone; showing, as antiquaries well know, that an age of bronze followed not an age of gold, but an age of stone.'

boats built of plank, and, as a modern shipwright would express himself, copper-fastened, cinerary urns,—for it would seem that, while in an earlier, as in a later age, our country-folk buried their dead, in this middle period they committed their bodies to the flames; and, withal, evidences, in the occasional productions of other countries, that commerce had begun to break up the death-like stagnation which characterized the earlier period, and to send through the nations its circulating tides, feeble of pulse and slow, but instinct, notwithstanding, with the first life of civilisation. And thus we reason on the same kind of unwritten data regarding the human inhabitants of our country who lived during these two early stages, as that on which we reason regarding their contemporaries the *extirpated* animals, or their predecessors the extinct ones. The interest which attaches to human history thus conducted on what may be termed the geologic plan is singularly great. No nation during its stone period possesses a literature; nor did any nation, of at least Western Europe, possess a literature during its bronze period. Of course, without letters there can be no history; and even if a detailed history of such uncivilized nations did exist, what would be its value? 'Milton did not scruple to declare,' says Hume, 'that the skirmishes of kites or crows as much merited a particular narrative as the confused transactions and battles of the Saxon Heptarchy.' But the subject arises at once in dignity and importance when, contemplating an ancient people through their remains, simply as men, we trace, step by step, the influence and character of their beliefs, their progress in the arts, the effects of invasion and conquest on both their minds and bodies, and, in short, the broad and general in their history, as opposed to the minute and the particular. The story of a civilized people I would fain study in the pages of their best and most philosophic historians; whereas I would prefer acquainting myself with that of a savage one archæologically and in its remains. And I would appeal, in justification of the preference, to the great superiority in interest and value of the recently published *Prehistoric Annals of Scotland*, by our accomplished townsman Mr. Daniel Wilson, over all the diffusive narrative and tedious description of all the old chroniclers that ever wore out life in cloister or cell.

What may be properly regarded as the geological deposits or formations of the two prehistoric periods in Scotland,— the period of stone and the period of bronze,—are morasses, sand dunes, old river estuaries, and that marginal strip of flat land which intervenes between the ancient and the existing coast lines. The remains of man also occur, widely scattered all over the country, in a superficial layer, composed in some localities of the drift-gravels, and in others of the boulder-clay; but to this stratum they do not *geologically* belong: they lie at a grave's depth, and have their place in it through the prevalence of that almost instinctive feeling which led the patriarch of old to bury his dead out of his sight. Most of the mistakes, however, which would antedate the existence of our species upon the earth, and make man contemporary with the older extinct mammals, have resulted from this ancient practice of inhumation, or from accidents which have arisen out of it.

All our Scotch morasses seem to be of comparatively modern origin. There are mosses in England, or at least buried forests, as on the Norfolk coast, at Cromer and Happisburgh, that are more ancient than the drift-clays and gravels; whereas, so far as is yet known, there are none of our Scotch mosses that do not *overlie* the drift formations; and not a few of their number seem to have been formed within even the historic ages. They are the memorials of a period, spread over many centuries, which began after Scotland had arisen out of the glacial ocean, and presented, under a softening climate, nearly the existing area, but bore, in its continuous covering of forest, the indubitable signs of a virgin country. It is remarked by Humboldt, that all the earlier seats of civilisation are bare and treeless. 'When, in passing from our thickly foliated forests of oak, we cross,' he says, 'the Alps or the Pyrenees, and enter Italy or Spain, or when the traveller first directs his eye to some of the African coasts of the Mediterranean, he may be easily led to adopt the erroneous inference that absence of trees is a characteristic of the warmer climates. But he forgets,' it is added, 'that Southern Europe wore a different aspect when it was first colonized by Pelasgian or Carthaginian settlers. He forgets, too, that an earlier civilisation of the human race sets bounds to the increase of forests; and that nations, in their change-loving spirit, gradually destroy the decorations which rejoice our eye in the north, and which, *more than the records of history, attest the youthfulness of our civilisation.*' Some of my audience must be old enough to remember the last of the great aboriginal woods of Scotland. It was only during the second war of the first French Revolution, when the northern ports of Europe were shut against Great Britain, that the native pine-woods of Rothiemurchus and the upper reaches of the Spey were cut down; and as late as the year 1820, I looked, in the upper recesses of Strathcarron, on the last scattered remains of one of the most celebrated of the old pine-forests of Ross-shire. Possibly some of the fragments of the pine-forest which skirted the western shores of Loch Maree may still exist; though, when I last passed through it, many years ago, the axe was busy among its glades. It is known of some of our Scotch mosses,—the deposits which testify geologically to this primitive state of things when the country was forest-covered,—that they date from the times of the Roman invasion, and were consequences of it. The mark of the Roman axe,—a narrow, chisel-like tool,—has been detected, in many instances, on the lower tier of stumps over which the peat has accumulated; and in some cases the sorely rusted axe itself has been found sticking in the buried tree. Among the tangled débris of a prostrated forest the woodman frequently mislays his tools,—a mishap to which the old Romans seem to have been as subject as the men of a later time; and so the list of Roman utensils, coins, and arms, found in the mosses of the south and midland parts of Scotland, is in consequence a long one. 'In Possil Moss, near Glasgow,' says Rennie, in his *Essay on Peat Moss*, 'a leathern bag containing about two hundred silver coins of Rome was found; in Dundaff Moor a number of similar coins were found; in Annan Moss, near the Roman Causeway, a Roman ornament of pure gold was found; a Roman camp-kettle was found eight feet deep under a moss on the estate of Ochtertyre; in Flanders Moss a similar utensil was found; a Roman jug was found in Lochar Moss, Dumfriesshire; a pot and decanter of Roman copper was found in a moss in Kirkmichael parish, in the same county; and two pair of vessels of Roman bronze in the Moss of Glenderhill, in Strathaven.' And thus the list runs on. It is not difficult to conceive how, in the circumstances, mosses come to be formed. The Roman soldiers cut down, in their march, wide avenues in the forests through which they passed. The felled wood was left to rot on the surface; small streams were choked up in the levels; pools formed in the hollows; the soil beneath, shut up from the light and the air, became unfitted to produce its former vegetation; but a new order of plants, the thick water-mosses, began to spring up; one generation budded and decayed over the ruins of another; and what had been an overturned forest became in the course of years a deep morass,—an unsightly but permanent monument of the formidable invader.

Some of our other Scotch mosses seem to have owed their origin to violent hurricanes;—their under tier of trunks, either torn up by the roots or broken across, lie all one way.

What may be termed their *native* fossils are exceedingly curious. I have seen personal ornaments of the stone period, chiefly beads of large size, made out of a pink-coloured carbonate of lime, which had been found in the bed of gravel on which one of our Galwegian mosses rested, and which intimated that the 'stone period' had commenced in the island ere this moss had begun to form. We find the same fact borne out by the Black Moss on the banks of the Etive, Argyleshire, where, under an accumulation of eight feet of peat, there occur irregularly oval pavements of stone, overlaid often by a layer of wood-ashes, and surrounded by portions of hazel stakes,—the remains, apparently, of such primitive huts as those in which, according to Gibbon, the ancient Germans resided, and which were, we are told, 'of a circular figure, built of rough timber, thatched with straw, and pierced at the top, to leave a free passage for the smoke.' Similar remains, but apparently of a still more ancient type, have been laid open in Aberdeenshire; and I find Mr. Wilson stating, in his archæological history, that on several occasions, rude canoes, which had been hollowed out of single logs of wood by the agency of fire, and evidently of the 'stone age,' have been found in Lochar Moss, Dumfriesshire, with ornamental torcs and brass bowls, not less evidently of the subsequent 'bronze period.' It is stated by Dr. Boate, in his *Natural History*, that in Ireland, the furrows of what had been once ploughed fields have been found underlying bogs,—in one instance at least (in Donegal), with the remains of an ancient plough, and the wattles of a hedge six feet beneath the surface. In 1833 there was discovered in Drumkilen bog, near the north-east coast of the county of Donegal, an ancient house formed of oak beams. Though only nine feet high, it consisted of two storeys, each about four feet in height. One side of the building was entirely open, and a stone chisel was found on the floor,—indicating that this ancient domicile belonged to the stone period. Associated, too, with the works of man of the earlier periods, we find in our mosses equally suggestive remains of the *extirpated*, and in some cases of the *extinct* animals, such as gigantic skulls and horns of the *Bos Primigenius* or native ox, and of the *Cervus Megaceros* or Irish elk, with the skeletons of wolves, of beavers, of wild horses, and of bears. There exists what seems to be sufficient evidence that the two extinct animals named the Irish elk and native ox were contemporary with the primitive hunters of the stone period : the cervical vertebræ of a native ox have been found deeply scarred by a stone javelin, and the rib of an Irish elk perforated by a stone arrow-head; and it is known that some of the extirpated animals, such as the wild horse, wolf, and beaver, continued to live among our forests down till a comparatively recent period.[1] We find it stated by Hector Boece in his History, that there were beavers living among our Highland glens even in his days, as late as the year 1526; but there rests a shadow of doubt on the statement. It is unquestionable, however, that the Gaelic name of the creature, *Lasleathin*, or broad-tail, still survives; and equally certain that when Baldwin, Archbishop of Canterbury, journeyed into Wales towards the close of the twelfth century, to incite the Welsh to join in the Crusades, the beaver was engaged in building its coffer domes and log-houses in the river Teivy, Cardiganshire. The wolf and wild horse maintained their place in at least the northern part of the island for several centuries later. When in 1618 Taylor, the water poet, visited Scotland, he accompanied the 'good lord of Mar' on one of his great hunting expeditions among the Grampians; and we find, from the

---

[1] Many interesting human remains have lately been disinterred from the Severn drift and gravels near Tewkesbury, such as cinerary urns with bones and ashes, and utensils for carrying water, associated with antlers of the red deer.—W. S. S.

amusing narrative of his journey, that for the space of twelve days he saw neither house nor corn-field, but deer, wild horses, wolves, and such like creatures. The wolf did not finally disappear from among our mountains until the year 1680, when the last of the race was killed in Lochaber by that formidable Ewan Cameron of Lochiel, with whom Cromwell was content to make peace after conquering all the rest of Scotland.

The sand dunes of the country,—accumulations of sand heaped over the soil by the winds, and in some cases, as in the neighbourhood of Stromness in Orkney, and near New Quay on the coast of Cornwall, consolidated into a kind of open-grained sandstone,—contain, like the mosses of the country, ancient human remains and works of art. There have been detected among the older sand dunes of Moray, broken or partially finished arrow-heads of flint, with splintered masses of the material out of which they had been fashioned,—the débris, apparently, of the workshop of some weapon-maker of the stone period. Among a tract of sand dunes on the shores of the Cromarty Firth, immediately under the Northern Sutor, in a hillock of blown sand, which was laid open about eighty years ago by the winds of a stormy winter, there was found a pile of the bones of various animals of the chase, and the horns of deer, mixed with the shells of molluscs of the edible species; and, judging from the remains of an ancient hill-fort in the neighbourhood, and from the circumstance that under an adjacent dune rude sepulchral urns were disinterred many years after, I have concluded that the hunters by whom they had been accumulated could not have flourished later than at least the age of bronze. It was ascertained in one of the Orkneys, about the year 1819, that a range of similar dunes, partially cleared by a long tract of high winds from the west, had overlain for untold ages what seemed to be the remains of an ancient Scandinavian village. In fine, very strange fossils of the human period has this sand deposit of subaërial formation been found to contain. There were disinterred on the Cornish coast in 1835, out of an immense wreath of sand, an old British church and oratory,—the church and oratory of Perran-sabulæ,—which had been hidden from the eye of man for nearly a thousand years. The Tractarian controversy had just begun at the time to agitate the Episcopacy of England; it had become of importance to ascertain the exact form of building sanctioned by antiquity as most conducive to devotion; and a fossil church, which had undergone no change almost since the times of the ancient Christianity, was too interesting a relic to escape the notice of the parties which the controversy divided. But though antagonistic volumes were written regarding it, in a style not quite like that in which Professor Owen and Dr. Mantell have since discussed the restoration of the Belemnite, it was ultimately found that the little old church of St. Pirran the Culdee,—such a building as Robinson Crusoe might have erected for the ecclesiastical uses of himself and his man Friday,—threw exceedingly little light on the vexed question of church architecture. The altar is in the east, said the Tractarians. Nay, the building itself does not lie east and west, replied their opponents. We grant you it does not, rejoined the Tractarians; but its gable fronts the point where the sun rises on the saint's birthday. Who knows that? exclaimed their opponents: besides, the sacred gable was unfurnished with a window. We deny that, said the Tractarians; the labourer who saw it just ere it fell says there was a large hole in it. And thus the controversy ran on, undoubtedly amusing, and, I daresay, very instructive. The north of Scotland has its ancient fossil barony underlying a wilderness of sand; ploughed fields and fences, with the walls of turf-cottages, and the remains of a manor-house, all irrecoverably submerged;—and we find the fact recorded in a

Scots Act of the times of William III. Curious, as being perhaps the only Act of Parliament in existence to which the geologist could refer for the history of a deposit, I must take the liberty of submitting to you a small portion of one of its long sentences. 'Our Sovereign Lord,' says the preamble, 'considering that many lands, meadows, and pasturages, lying on the sea-coasts, have been ruined and overspread in many parts of this kingdom by sand driven from sand-hills, the which has been mainly occasioned by the pulling up of the roots of bent, juniper, and broom bushes, which did loose and break the surface and scroof of the sand-hills; and particularly, considering that the barony of Cawbin, and house and yeards thereof, lying within the sheriffdom of Elgin, is quite ruined and overspread with sand, the which was occasioned by the foresaid bad practice of pulling the bent and juniper,—does hereby strictly prohibit,' etc. etc. etc. I have wandered for hours amid the sand-wastes of this ruined barony, and seen only a few stunted bushes of broom, and a few scattered tufts of withered bent, occupying, amid utter barrenness, the place of what, in the middle of the seventeenth century, had been the richest fields of the rich province of Moray; and, where the winds had hollowed out the sand, I have detected, uncovered for a few yards'-breadth, portions of the buried furrows, sorely dried into the consistence of sun-burned brick, but largely charged with the seeds of the common corn-field weeds of the country, that, as ascertained by experiment by the late Sir Thomas Dick Lauder, still retain their vitality. It is said that an antique dove-cot in front of the huge sand-wreath which enveloped the manor-house, continued to present the top of its peaked roof over the sand, as a foundered vessel sometimes exhibits its vane over the waves, until the year 1760. The traditions of the district testify that, for many years after the orchard had been enveloped, the topmost branches of the fruit-trees, barely seen over the surface, continued each spring languidly to throw out bud and blossom ; and it is a curious circumstance, that in the neighbouring churchyard of Dike there is a sepulchral monument of the Culbin family, which, though it does not date beyond the reign of James VI., was erected by a lord and lady of the last barony, at a time when they seem to have had no suspicion of the utter ruin which was coming on their house. The quaint inscription runs as follows :—

VALER : KINNAIRD : ELIZABETH : INNES : 1613 :
THE : BVILDARS : OF : THIS : BED : OF : STANE :
AR : LAIRD : AND : LADIE : OF : COVBINE :
QVHILK : TVA : AND : THARS : QVHANE : BRAITHE IS : GANE :
PLEIS : GOD : VIL : SLEIP : THIS : BED : VITHIN :

I refer to these facts, though they belong certainly to no very remote age in the past history of our country, chiefly to show that in what may be termed the geological formations of the human period very curious fossils may be already deposited, awaiting the researches of the future. As we now find, in raising blocks of stone from the quarry, water-rippled surfaces lying beneath, fretted by the tracks of ancient birds and reptiles, there is a time coming when, under thick beds of stone, there may be detected fields and orchards, cottages, manor-houses, and churches,—the memorials of nations that have perished, and of a condition of things and a stage of society that have for ever passed away.

Sand dunes and morasses are phenomena of a strictly local character. The last great geological change, general in its extent and effects, of which Scotland was the subject, was a change in its level, in relation to that of the ocean, of from fifteen to thirty feet. At some unascertained period, regarded as recent by the geologist—for man seems to have been an actor on the scene at the time,—but remote by the historian—for its date is anterior to that of his oldest authorities in this country,—the land rose, apparently during several interrupted paroxysms of upheaval, so that there was a fringe of comparatively level sea-bottom laid dry, and added to the country's area, considerably broader than that which we now see exposed by the ebb of every stream tide. And what I must deem indubitable marks of this change of level can be traced all around Scotland and its islands. The country, save in a few interrupted tracts of precipitous coast, where the depth of the water, like that beside a steep mole whose base never dries at ebb, precluded any accession to the land, presents around its margin a double coast line,— the line at present washed by the waves, and a line now covered with grass, or waving with shrubs, or skirted by walls of precipice perforated with caves, against which the surf broke for the last time more than two thousand years ago. These raised beaches form a peculiar feature in our Scottish scenery, which you must have often remarked. In passing along the public road between Portobello and Leith, the traveller sees upon the left hand a continuous grassy bank, with a line of willows atop, which he may mark in some places advancing in low promontories, in others receding into shallow bays, and which is separated from the present coast line, which in general flatness it greatly resembles, by a strip of rich meadow land, varying from one to three hundred yards in breadth. That continuous grassy bank is the old coast line; and the gently sloping margin of green meadow is the strip of flat sea-beach along which the tides used to rise and fall twice every twenty-four hours, ere the retreat of the sea within its present bounds. Should it be low ebb at the time, one may pass from the ancient to the recent sea-beach; the one waving with grass, the other brown with algæ; the one consisting, under its covering of vegetable mould, of stratified gravels and sand, blent with the decayed shells of mollusca that died more than twenty centuries since,—the other formed of exactly the same sort of lines of stratified sand and gravel, and strewed over by shells that were thrown ashore by the last tide, and that lived only a few weeks ago. And, rising over the lower, as over the upper flat, we see a continuous escarpment, which marks where, in the present age, during the height of stream tides, the sea and the land meet; just as the upper willow-crested escarpment indicates where they met of old. The two escarpments and the two gently sloping planes at their base are repetitions of the same phenomena, save that the upper escarpment and upper plane are somewhat softer in their outline than the lower,—an effect of the wear of the elements, and of the accumulation of the vegetable mould. There is as thorough an identity between them as between two contiguous steps of a stair, covered, the one by a patch of brown, and the other by a patch of green, in the pattern of the stair-carpet. There are other parts of our Scottish shores in which the old coast line is of a much bolder character than anywhere in this neighbourhood, and the plane at its base of greater breadth. On the Forfarshire coast, the Dundee and Arbroath Railway runs along the level margin, once a sea-bottom, which at one point, opposite the parish church of Barry, is at least two miles in breadth, and the old coast line rises from thirty to fifty feet over it. It is strongly marked on the southern side of the Dornoch Firth, immediately below and for several miles to the east of the town of Tain, where it attains a breadth of from one to two miles, and where the old sea-margin, rising over the cottage-mottled plain below in a series of jutting headlands, with green bosky bays between, strikes even the least practised eye as possessed of all the characteristic peculiarities of a true coast line. It is scarce less marked in the neighbourhood of Cromarty, and on the opposite shores of the Cromarty Firth, in the parish of Nigg. It runs along by much the greater portion of the eastern coast of Sutherland; and forms at the head of Loch Fleet, in the neighbourhood of Dornoch, a long withdrawing firth, bounded by picturesque shores, and covered by a short, green sward, level as the sea in a calm, on which groups of willow and alder trees take the place of busy fleets, and the hare and the partridge that of the coot and the porpoise. Along the upper recesses of almost all our flatter firths, such as the firths of Beauly, of Dingwall, of the Tay, and of the Clyde, it exists as fertile tracts of carse-land; the rich links of the Forth, rendered classical by the muse of Macneil, belong to it; it furnishes, in various other localities more exposed to the open sea, ranges of sandy links of a less valuable character, such as the range in our own neighbourhood occupied by the race-course of Inveresk; and not a few of the seaports and watering-places of the country, such as the greater part of Leith, Portobello, Musselburgh, Kirkcaldy, Dundee, Dingwall, Invergordon, Cromarty, Wick, Thurso, Kirkwall, Oban, and Greenock, have been built upon it.

The old coast line, with the flat marginal selvage at its base, form, as I have said, well-marked features in the scenery of the island. Geology may be properly regarded as the *science* of landscape : it is to the landscape-painter what anatomy is to the historic one or to the sculptor. In the singularly rich and variously compounded prospects of our country there is scarce a single trait that cannot be resolved into some geological peculiarity in the country's framework, or which does not bear witness otherwise and more directly than from any mere suggestion of the associative faculty, to some striking event in its physical history. Its landscapes are tablets roughened, like the tablets of Nineveh, with the records of the past; and their various features, whether of hill or valley, terrace or escarpment, form the bold and graceful characters in which the narrative is inscribed. As our Scottish geologists have given less attention to this special department of their science than to perhaps any other,—less, I am disposed to think, than, from its intrinsic interest and its bearing on art, is fairly owing to it,—I shall take the liberty—casting myself on the forbearance of such of my audience as are least artistic in their tastes—of occasionally touching upon it in my course.

I need scarce refer to the scenery of our mosses,—these sombre, lake-like tracts, divested, however, of the cheerful gleam of the water,—that so often fatigue the eye of the traveller among our mountains, but which at that season when the white cottony *carnach* mottles their dark surfaces, reminding one of tears on a hatchment,—when the hills around, purple with the richly-blossoming heath, are chequered with the light and shade of a cloud-dappled sky,—and when, in the rough foreground, the grey upright stone of other days waves its beard of long grey lichen to the breeze, —are not unworthy, in their impressive loneliness, of employing, as they have oftener than once done, the magic pencil of a Macculloch. I need as little refer to the scenery of those sand dunes which gleam so brightly amid some of our northern landscapes, and which, not only in colour, but also in form, contrast so strongly with our morasses. The dark flat morass is suggestive always of sluggish and stagnant repose; whereas among our sand dunes, from the minuter ripple-markings of the general surface, to the wave-like form of the hills sloped in the direction of the prevailing winds, and curved, like snow-wreaths, to the opposite point of the compass, almost every outline is equally suggestive of motion. I could, however, fain borrow the pencil of our countryman Hill, as he employs it in his exquisite cabinet-pictures, to portray the story of the last Barony: rolling hills of sand all around, the red light of a stormy summer evening deepening into dun and lurid brown, through an eddying column of suffocating dust snatched up by a whirlwind; the antique garden-dial dimly shadowing forth the hour of sunset for the last time amid half-submerged shrubs and trees; and, full in the centre of the picture, a forlorn fortalice of the olden time, with the encroaching wreath rising to its lower battlements, like some wrecked vessel on a wild lee-shore, with the angry surf raging high over her deck, and kissing with its flame-like tips the distant yards.

The scenery of the old coast line possesses well-nigh all the variety of that of the existing coast; but it substitutes field and meadow for the blue sea, and woods and human dwellings for busy mast-crowded harbours, and fleets riding at anchor. It is pleasing, however, to see headland jutting out beyond headland into some rich plain, traversed by trim hedgerows and green lanes; or some picturesque cottage, overshadowed by its gnarled elm, rising in some bosky hollow at the foot of the swelling bank or weather-stained precipice, beneath which the restless surf once broke against the beach. There are well-marked specimens of this scenery of the ancient coast line in our immediate neighbourhood. Musselburgh, with its homely Saxon name, lies in the middle of what was once a flat sandy bay, now laid out into fields, gardens, and a race-course; and the old coast escarpment, luxuriant with hanging woods, and gay with villas, and which may possibly have been its first Celtic designation, *Inveresk*, ere the last upheaval of the land, half-closes around it. The church and burying-ground occupy the top of a long ridge, that had once been a river-bar, heaped up apparently by the action of the waves on the one side, and by that of the stream on the other. But, as shown by the remains of Roman baths and a Roman rampart, which once occupied its summit, it must have borne its present character from at least the times of Lollius Urbicus,—perhaps for several centuries earlier. The neighbouring town of Portobello, as seen from the east, just as it comes full in sight on the Musselburgh road, seems set so completely in a framework of the ancient escarpment, that it derives from it all its natural features. But it is where, along our bolder shores, lines of steep precipices have been elevated over the sea, so that the waves no longer reach their bases, that the old coast scenery is at once most striking and peculiar. Tall picturesque stacks, which had once stood up amid the surf, brown and shaggy with the serrated fucus and the broad-fronded laminaria, now rise out of thickets of fern or sloethorn, and wave green with glassy ivy and the pendant honeysuckle. Deep caverns, too, in which the billows had toiled for ages, but now silent, save when the drop tinkles from above into some cool cistern half-hidden in the gloom of the interior, open along the wall of cliffs; and over projecting buttresses of rock, perforated often at their bases as if by Gothic archways, and thickly mantled over by liver-worts, green and grey, the birch hangs tremulous from above, or the hazel shoots out its boughs of brighter green, or the mountain-ash hangs its scarlet berries. One of the most pleasing landscapes of one of the most accomplished of female artists,—Miss Stoddart,—has as its subject an ancient escarpment of this bold character, which occurs in Arran. A mossy, fern-tufted meadow, skirted by the sea, roughened by what had once been half-tide skerries, and enlivened by a Highland cottage, stretches out into the foreground from an irregular wall of rock, overhung by graceful foliage, hollowed into deep recesses, adown which the waters trickle, and with some of its bolder projections perforated at the base like flying buttresses of the decorated Gothic; and such is the truth of the representation, that we at once determine that the artist had chosen as her subject one of the more precipitous reaches of the old coast line, and that its wall of rock must have derived much of the peculiarity of trait so happily caught, from the action of the waves. Again, in direct contrast with this striking type of old coast escarpment, though in its own way not less striking, Mr. Hill's fine picture, 'The Sands at Sunrise,' lately engraved by the Art Union, exhibits as its background one of those long, flat, sandy spits, products of the last upheaval, which, stretching far into the sea, bear amid the light of day an air of even deeper loneliness than our woods and fields when embrowned by the gathering night. When the insulated stacks of an old coast line are at once tall and attenuated, and of a white or pale-coloured rock, the effect, especially when viewed by moonlight, is singularly striking. The valley of the Seine, as described by Sir Charles Lyell,—now a valley, but once a broad firth,—is flanked on each side, in its lower reaches, by tall stacks of white chalk, of apparently the same age as those of the ancient coast line of our own country; and, seen ranged along their green hill-sides, in the imperfect light of evening, or by the rising moon, they seem the sheeted spectres of some extinct tribe of giants.

The date of that change of level which gave to Scotland this flat fringe of margin-land, with its picturesque escarpment of ancient coast, we cannot positively fix. We find reason to conclude that it took place previous to the age of the Roman invasion. It has been shown, from evidence of a semi-geologic, semi-archæologic character, by one of our highest authorities on the subject, Mr. Smith of Jordanhill, that the land must have stood at a not lower level than now, when the Roman wall which connects the firths of Forth and Clyde was completed. For, had it been otherwise, some of the terminal works which remain would have been, what they obviously were not, under the sea line at the time. In the sister kingdom, too, which has also its old coast line, St. Michael's Mount in Cornwall, which was connected with the mainland at low water by a strip of beach in the times of Julius Cæsar,—a fact recorded by Diodorus Siculus,—is similarly connected with the mainland at low water still. But though the upheaval of the old coast line is removed thus beyond the historic period, it seems to have fallen, as I have said, within the human one : man seems to have been an inhabitant of the island when its general level was from twenty to forty feet lower than now, and the waves broke at full tide against the old coast line. 'The skeleton of a Balænoptera,' says Professor Owen, 'seventy-two feet in length, was found,' about thirty years ago, 'imbedded in the clay on the banks of the Forth, more than twenty feet above the reach of the highest tide.' And again, 'Several bones of a whale,'[1] he continues, 'were also discovered at Dunmore rock, Stirlingshire, in brick-earth, nearly forty feet above the present sea-level.' These whales must have been stranded when the old coast line was washed by the waves, and the marginal strip existed as an oozy sea-bottom; and yet in both cases there were found among the bones primitive weapons made of the pointed branches of deer's horns, hollowed at their broad ends by artificial perforations; and in one of these perforations the decayed fragments of a wooden shaft still remained. The pointed and perforated pieces of horn were evidently rude lance-heads, that in all probability had been employed against the stranded *cetacea* by the savage natives. Further, where the city of Glasgow now stands, three ancient boats,—one of which may be seen in the Museum of our Scottish Antiquaries in Edinburgh, and another in the Andersonian Museum,—have been dug up since the year 1781; the last only four years ago. One of the number was found a full quarter of a mile from the Clyde, and about twenty-six feet above its level at high water. It reposed, too, not on a laminated silt, such as the river now deposits, but on a pure sea-sand. 'It therefore appears,' says Mr. Robert Chambers, in his singularly ingenious work on *Raised Beaches*, 'that we have scarcely an alternative to the supposition that when these vessels foundered, and were deposited where in modern times they have been found, the

---

[1] Bones of the whale have been found in the clay of the Avon and Severn drifts, in a similar position.—W. S. S.

Firth of Clyde was a sea several miles wide at Glasgow, covering the site of the lower districts of the city, and receiving the waters of the river not lower than Bothwell Bridge.' I may add, that the Glasgow boat in the Antiquarian Museum is such a rude canoe, hollowed out of a single trunk, as may be seen in use among such of the Polynesian islands as lie most out of the reach of civilisation, or in the Indian Archipelago, among the rude Alforian races; and that in another of these boats,—the first discovered,—there was found a beautifully polished hatchet of dark greenstone,—an unequivocal indication that they belonged to the 'stone period.' There are curious etymologies traceable among the older Celtic names of places in the country which I have sometimes heard adduced in evidence that it was inhabited, ere the last upheaval of the land, by the ancient Gaelic-speaking race. Eminences that rise in the flat marginal strip, and which, though islands once, could not have been such since the final recession of the sea, continue to bear, as in the neighbourhood of Stirling, the Gaelic prefix for an island. But as the old Celts seem to have been remarkable as a people for their nice perception of resemblances, the insular form of these eminences may be perhaps regarded as suggestive enough to account for their names. One of these etymologies, however, which could scarce have been founded on any mere resemblance, seems worthy of special notice. Loch Ewe, in Ross-shire, one of our salt-sea lochs, receives the waters of Loch Maree,—a noble fresh-water lake, about eighteen miles in length, so little raised above the sea-level, that ere the last upheaval of the land it must have formed merely the upper reaches of Loch Ewe. The name Loch Maree,—Mary's Loch,—is evidently mediæval. And, curiously enough, about a mile beyond its upper end, just where Loch Ewe would have terminated ere the land last arose, an ancient farm has borne from time immemorial the name of Kinlochewe,—the head of Loch Ewe.

Dispose, however, of the etymologies as we may, there are facts enough on record which render it more than probable that, though the general change of level to which we owe the old coast line in Scotland does not lie within the historic ages, it is comprised within the human period. But we cannot, as has been shown, fix upon a date for the event.

Were the case otherwise,—could we fix with any certainty the time when this change of level took place, and the platform of the lower coast line was gained from the sea,—there might be an approximation made to the anterior space of time during which the line of high water had been the willow-crowned escarpment beyond Portobello and the green bank near Rutherglen, and the sea rose far beyond its present limits in our firths and bays. There are portions of the coast that at this early period presented to the waves lines of precipices that are now fringed at their bases by strips of verdure, and removed far beyond their reach. There are other portions of coast in the immediate neighbourhood of these, where similar lines of precipices, identical in their powers of resistance, were brought by the same movement within that very influence of the waves beyond which the others had been raised. And each line bears, in the caves with which it is fretted,—caves hollowed by the attrition of the surf in the direction of faults, or where masses of yielding texture had been included in the solid rock,—indices to mark, proportionally at least, the respective periods during which they were exposed to the excavating agent. Thus, the average depth of the ancient caves in an exposed line of coast, as ascertained by dividing the aggregate sum of their depths by their number, and the average depth, ascertained by the same process, of the recent caves, equally exposed on the same coast, and hollowed in the same variety of rock, could scarce fail to represent their respective periods of exposure, had we but a given number of years, historically determined, to set off against the average measurement of the recent excavations. Even wanting that, however, it is something to know, that though the sea has stood at the existing sea-margin since the days of Agricola, and at least a few centuries more, it stood for a considerably longer period at the old coast line. The rock of which those remarkable promontories, the Sutors of Cromarty, are composed, is a granitic gneiss, much traversed by faults, and enclosing occasional masses of a soft chloritic schist, that yields to the waves, while the surrounding gneiss,—hard enough to strike fire with steel,—remains little affected by the attrition of centuries. These promontories have, in consequence, their numerous caves ranged in a double row,—the lower row that of the existing coast, the upper that of the old one; and I have examined both rows with some little degree of care. The deepest of the recent caves measures, from the opening to its inner extremity, where the rock closes, exactly a hundred feet; the deepest of the ancient ones, now so completely raised above the surf, that in the highest tides, and urged upwards by the severest storms, the waves never reach its mouth, measures exactly a hundred and fifty feet. And these depths, though much beyond the respective average depths of their several rows, bear, so far as I could ascertain the point, the proportions to each other that these averages bear. The caves of the existing coast line are as *two* in depth, and those of the old coast line as *three*. If the excavation of the recent caves be the work of *two* thousand years, the excavation of the ancient caves must have been the work of *three* thousand; or, as two thousand does not bring us much beyond the Roman period, let us assume as the period of the existing coast line and its caves, two thousand two hundred years, and as the proportional period of the old coast line, three thousand three hundred more. Both sums united bring us back five thousand five hundred years. How much more ancient either coast line may be, we of course cannot determine:

we only know that, on the lowest possible assumption, we reach a period represented by their united ages only less extended by six years than that which the Samaritan chronology assumes as the period during which man has existed upon earth, and only three hundred and fifty-five years less than that assumed by the Masoretic chronology. The chronology of the Septuagint, which many have begun to deem the most adequate of the three, adds about five hundred and eight-six years to the sum of the latter.

Permit me, in closing this part of my subject, to show you that changes of level such as that to which we owe our old coast line in Scotland, and the marginal strip of dry land which we have laid out into so many pleasant gardens and fields, and on which we have built so many of our seaport towns, are by no means very rare events to the geologist. He enumerates at least five localities in the Old World,—Scandinavia, part of the west coast of Italy, the coasts of Cutch and of Arracan, and part of the kingdom of Luzan, in which the level is slowly changing at the present time; and in the New World there are vast districts in which the land suddenly changed its level for a higher one during the present century. 'On the 19th of November 1822,' says Sir Charles Lyell, 'the coast of Chili was visited by a most disastrous earthquake. When the district around Valparaiso was examined on the morning after the shock, it was found that the whole line of coast for the distance of above one hundred miles was raised above its former level. At Valparaiso the elevation was three feet, and at Quinteno about four feet. Part of the bed of the sea remained bare and dry at high water, with beds of oyster, mussel, and other shells, adhering to the rocks on which they grew,—the fish being all dead, and exhaling offensive effluvia.' Again, on the east side of the Bay of Bengal, upon the coast of Arracan, which is at present in the course of rising, there are islands which present on their shores exactly such an appearance as our own country would have presented some sixty or a hundred years after the elevation of the old coast line. The island of Reguain, one of these, was carefully surveyed in the year 1841 by the officers of Her Majesty's brig Childers; and it has been carefully mapped in the admirable Physical Atlas of the Messrs. Johnston of Edinburgh. We find it, as shown in the map, resembling three islands; the one placed within the other, as, to employ a homely illustration, the druggist, to save room, places his empty pill-boxes the one within the other. First, in the centre, there is the ancient island, with a well-defined coast line, some six or eight feet high, running all around it. At the base of this line there is a level sea of rich paddy fields,—for what may be termed the second island has been all brought into cultivation; and *it* has also its coast line, which descends some six or eight feet more, to the level of a third island, which was elevated over the sea not more than eighty years ago, and which is still uncultivated; and the third island is surrounded by the existing coast line. Thus the centre island of Reguain consists of three great steps or platforms, each of which marks a paroxysm of elevation; and, with the upheaval of the coast of Chili, and a numerous class of events of a similar character, it enables us to conceive of the last great geological change of which our country was the subject. We imagine a forest-covered land, marked by the bold commanding features by which we recognise our country, but inhabited by barbarous, half-naked tribes, that dwell in rude circular wigwams, formed of the branches of trees,— that employ in war or the chase weapons of flint or jasper, —and that navigate their rivers or estuaries in canoes hollowed by fire out of single logs of wood. There has been an earthquake during the night; and when morning rises, the beach shows its broad darkened strip of apparent ebb, though the tide is at full at the time; and when the waters retire, they leave uncovered vast tracts never seen before, comparatively barren in sea-weed, but rich in stony nulliporite incrustations, minute corallines, and fleshy sponges. Ages elapse, and civilisation grows. The added belt of level land is occupied to its utmost extent by man : he lays it out into gardens and fields, and builds himself a dwelling upon it : but no sooner has he rendered it of some value, than the sea commences with him a course of tedious litigation for the recovery of its property ; and bit by bit has it been wrested out of his hands. Almost all those tracts on our coasts which have been suffering during the last few centuries from the encroachment of the waves, and which have to be protected against their fury wherever land is valuable, as in this neighbourhood, by lines of bulwarks, belong to the flat marginal strip won from them by the last change of level.

Our next great incident in the geologic history of Scotland dates, it would seem, beyond the human period. In passing along the beach between Musselburgh and Portobello, or again between Portobello and Leith, or yet again between Leith and Newhaven, one sees an exceedingly stiff, dark-coloured clay, charged with rounded pebbles and boulders, and which, where washed by the waves, presents a frontage nearly as steep as that of the rock itself. The deposit by which it is represented is known technically to the agriculturist as *Till*, and to the geologist as the *Boulder-Clay*. Though not continuous, it is of very general occurrence, in the Lowlands of Scotland, and presents, though it varies in colour and composition, according to the nature of the rocks which it overlies, certain unique appearances, which seem to connect its origin in the several localities with one set of causes, and which no other deposit presents. Like the raised beaches, it has contributed its distinctive quota to the variously featured scenery of our country. The Scottish word *scaur*, in the restricted significancy attached to it in many parts of the kingdom, means simply a precipice of clay, and it is almost invariably the boulder-clay that forms *scaurs* in Scotland; for it is one of the peculiarities of the deposit, that it stands up well-nigh as steeply over the sides of rivers, or on encroaching sea-beaches, or on abrupt hill-sides, as rock itself; and these clay precipices bear almost invariably a peculiar set of characters of their own. In some cases they spring up as square and mural, seen in front, as cliffs of the chalk, but seen in profile, we find their outlines described by parabolic curves. In other cases we see the vegetable mould rendered coherent by the roots of shrubs and grasses projecting over them atop, like the cornice of some edifice over its frieze. In yet other cases, though abrupt as precipices of solid rock, we find them seamed by the weather into numerous divergent channels, with pyramidal peaks between; and, thus combining the perpendicularity of true cliffs with the rain-scooped furrows of a yielding soil, they present eccentricities of aspect which strike, by their grotesqueness, eyes little accustomed to detect the picturesque in landscape. Such are some of the features of the *scaurs* of our country,—a well-marked class of precipices for which the English language has no name. It is, however, in continuous grass-covered escarpments, which in some parts form the old coast-line, and rise in others along the sides of rivers, that we detect at once the most marked and most graceful scenic peculiarity of the boulder-clay. The steep slopes, furrowed by enormous flutings, like those of the antique Doric, appear as if laid out into such burial-mounds as those with which a sexton frets the surface of a country churchyard, but with this difference, that they seem the burial-mounds of giants tall and bulky as those that of old warred against the gods. On a grass-covered escarpment of the boulder-clay in the neighbourhood of Cromarty, these mounds are striking enough to have caught the eye of the children of the place, and are known among them as the giants' graves. They lie against the green bank, each from forty to sixty yards in length, and from six to ten yards in height, with their feet to the shore, and their heads on the top of the escarpment; and when the evening sun falls low, and the shadows lengthen, they form, from their alternate bars of light and shade, that remind one of the ebon and ivory buttresses of the poet, a singularly pleasing feature in the landscape. I have sometimes wished I could fix their features in a calotype, for the special benefit of my friends the landscape painters. This vignette, I would fain say, represents the boulder-clay after its precipitous banks— worn down, by the frosts and rains of centuries, into parallel runnels, that gradually widened into these hollow grooves—had sunk into the angle of inclination at which the disintegrating agents ceased to operate, and the green sward covered all up. You must be studying these peculiarities of aspect more than ever you studied them before. There is a time coming when the connoisseur will as rigidly demand the specific character of the various geologic deposits in your rocks and *scaurs*, as he now demands specific character in your shrubs and trees.

I have said that the boulder-clay exhibits certain unique appearances, which connect its origin in the several localities with one set of causes, and which no other deposit presents. On examining the boulders which it encloses, we find them strongly scarred and scratched. In most instances, too, the rock on which the clay rests,—if it be a trap, or a limestone, or a finely-grained sandstone, or, in short, any rock on which a tool could act, and of a texture fitted to retain the mark of the tool,—we find similarly scarred, grooved, and scratched. In this part of the country the boulder-clay contains scarce any fossils, save fragments of the older organisms derived from the rocks beneath; but in both the north and south of Scotland,—in

Caithness, for instance, and in Wigtonshire,—it contains numerous shells, which, both in their species and their state of keeping, throw light on the origin of the formation. But of that more anon. Let me first remark, that the materials of the level marginal strip of ancient sea-beach beneath the old coast line seem, like the materials of the existing sea-beach, to have been arranged wholly by the agency of water. But in the boulder-clay we find a class of appearances which mere water could not have produced. Not only are the larger pebbles and boulders of the deposit scratched and grooved, but also its smaller stones, of from a few pounds to but a few ounces, or even less than an ounce, in weight; and this, too, in a peculiar style and direction. When the stones are decidedly of an oblong or spindle shape, the scratchings occur, in at least four cases out of every five, in the line of their longer axis. Now, the agent which produced such effects could not have been simply water, whether impelled by currents or in waves. The blacksmith, let him use what strength of arm he may, cannot bring his file to bear upon a minute pin until he has first locked it fast in his vice; and then, though not before, his tool bears upon it, and scratches it as deeply as if it were a beam of iron of a ton weight. The smaller stones must have been fastened before they could have been scratched. Even, however, if the force of water could have scratched and furrowed them, it would not have scratched and furrowed them longitudinally, but across. Stones, when carried adown a stream by the torrent, or propelled upwards along a beach by the waves, present always their broader and longer surfaces; and the broader and longer these surfaces are, the further are the stones propelled. They are not *launched* forwards, as a sailor would say, *end on*, but *tumbled* forwards *broadside*. They come rolling down a river in flood, or upwards on the shore in a time of tempest, as a hogshead rolls down a declivity.

In the boulder-clay, on the contrary, most of the pebbles that bear the mark of their transport at all were not *rolled*, but *slidden* forward in the line of their longer axis. They were launched, as ships are launched, in the line of least resistance, or as an arrow or javelin is sent on its course through the air. Water could not have been the agent here, nor yet an eruption of mud propelled along the surface by some wave of translation produced by the sudden upheaval of the bottom of the sea, or by some great wave raised by an earthquake.

But if water or an eruption of mud could not have produced such effects as the longitudinal scratching, let us ask what could have produced them? There are various processes going on around us, by which the scratchings on the solid rocks beneath are occasionally simulated with a less or greater degree of exactness. In some of our shallow Highland fields, for instance, I have seen the rock beneath, or the stones buried at the depth of but a few inches from the surface, scarred by the plough with ruts not very unlike the larger ones on the stones and rocks of the boulder-clay; but in these plough-scarred surfaces the polish is wanting. Again, in some of our steeper lanes, if a fine-grained trap has been used in the pavement, we find that it soon polishes and wears down under the iron-armed feet of the passengers, and becomes scratched in the line of their tread, in a style not very distinguishable, save for the absence of the deeper furrows, from that of the scratched and polished rock-pavements of the boulder-clay. But I know of only one process by which, on a small scale, *all* the phenomena of the boulder-clay could be produced,—more especially, however, the phenomena of its oblong pebbles scratched in the lines of their longer axis; and my recollection of that one dates a good many years back. When, more than a quarter of a century ago, the herring fishing began to be prosecuted with vigour in the north of Scotland, many of the Highland woods of natural birch and alder were cut down for the manufacture of barrels, and floated in rafts along the rivers to the sea. And my opportunities of observing these rafts, as they shot along the more rapid reaches of our mountain streams, or swept over their shallower ledges, grazing the bottom as they passed, naturally led me to inquire into their operations upon the beds of the streams adown which they were floated. Let us advert to some of these. When a large raft of wood, floated down a rapid river, grates heavily over some shallow bank of gravel and pebbles resting on the rock beneath, it communicates motion, not of the *rolling* but of the *lurching* character, to the flatter stones with which it comes in contact. It slides ponderously over them; and they with a speed diminished in ratio from that of the moving power in proportion to the degree of friction below or around, slide over the stones or rock immediately beneath. And thus, to borrow my terminology from our Scotch law courts, they are converted at once into *scratchers* and *scratchees*. They are scratched by the grating, sand-armed raft, which of course moves quicker than they move; and they scratch, in turn, the solid mass or embedded fragment along which they are launched. Further, if the gravelly shoals of the stream have, as is not uncommon in the shallows of our Highland rivers, their thickly-set patches of pearl mussels, many of these could scarce miss being crushed and broken; and we would find not a few of their fragments, if much subjected to the friction of the rafting process, rounded at their edges, and mayhap scratched and polished like the stones. Nor is it difficult to conceive of a yet further consequence of the process. A vast number of rafts dropping down some river from day to day and year to year, and always grating along the same ledges of sandstone, trap, or shale, would at length very considerably wear them down; and the materials of the waste, more or less argillaceous, according to the quality of the rock, would be depo-

C

sited by the current in the pools and gentler reaches of the stream below. Even the continual tread of human feet in a crowded thoroughfare soon wears down the trap or sandstone pavement, and converts the solid stone into impalpable mud. Further, the colour of the mud or clay would correspond, as in the thoroughfare or public road, with the colour of the rocks or stones which had been grooved down to form it; and there would occasionally mingle in the mass thus originated, rounded fragments of shells and pebbles scratched in the line of their longer axis.

Now, in the boulder-clay we find all these peculiarities remarkably exemplified. It contains, as has been shown, the oblong stones scratched longitudinally; we find it thickly charged in various parts of Scotland, though not in our own immediate neighbourhood, with worn and rounded fragments of broken shells; and we see it almost invariably borrowing its colour from the rocks on which it rests,—a consequence, apparently, of its being the dressings of these rocks. There is a peculiar kind of clay which forms on the surface of a hearthstone or piece of pavement, under the hands of a mason's labourer engaged in rubbing it smooth with water and a polisher of gritty sandstone. This clay varies in quality and colour with the character of the stone operated upon. A flag of Arbroath pavement yields a bluish-coloured clay; a flag of the Old Red of Ross or Forfarshire, a reddish coloured clay; a flag of Sutherlandshire Oolite, or of the Upper Old Red of Moray or of Fife, a pale yellowish clay. The polishing process is a process which produces clay out of stones as various in tint as the colouring of the various stones which yield it; and in almost every instance does the clay thus formed resemble some known variety of the boulder-clay. The boulder-clay, in the great majority of cases, is, both in colour and quality, just such a clay as might be produced by this recipe of the mason's labourer from the rocks on which it rests. The red sandstone rocks of Moray, Cromarty, and Ross are covered by red boulder-clays; a similar red boulder-clay overlies the red sandstone rocks of Forfarshire; and I was first apprised, when travelling in Banffshire some years ago, that I had entered on the district of the Old Red, by finding the boulder-clay assuming the familiar brick-red hue. Over the pale Oolites of Sutherlandshire, as at Brora and Golspie, it is of a pale yellow tint, and of a yellowish red over the pale Old Red Sandstones of the long flat valley known as the *Howe* of Fife. Again, in the middle and north-western districts of Caithness, where the paving flagstones so well known in commerce give to the prevailing rocks of the district a sombre tint of grey, the boulder-clay assumes, as in the neighbourhood of Wick and Thurso, the leaden colour of the beds which it overlies; while over the Coal Measures of the south of Scotland, as in East and West Lothian, and around Edinburgh, it is of a bluish-black tint,—exactly the colour which might be premised, on the polishing theory, from the large mixture of shale-beds, coal-seams, and trap-rocks, which occurs amid the prevailing light-hued sandstones of the deposits beneath. Of course, this condition of resemblance in average colour between the rocks and the boulder clays of a district is but of *general*, not *invariable*, occurrence,—the boulder-clay is not *invariably* the dressings of the rocks beneath. We may occasionally find the trail of the rubbings of one tract overlying, in an easterly direction, the deposits of a different one; just as we would find the rubbings of variously-coloured pieces of pavement laid down to form a floor, and then polished square by square where they lay, encroaching, the débris of one square on the limits of another, in the direction of the outward stroke of the polisher.

But while we thus find all the conditions of a raft-formed deposit *in* or *associated with* the boulder-clay,—such as grooved and furrowed rocks beneath, scratched and polished stones, lined longitudinally, enclosed in it, accompanied, in not a few instances, by rounded fragments of shells, and a general conformity in its colour to that of the rocks on which it rests,—where in nature shall we find the analogues of the producing rafts themselves? A native of Newfoundland, who season after season had seen the Arctic icebergs grating heavily along the coasts of the island, would experience little difficulty in solving the riddle. For rafts of wood we have but to substitute rafts of ice, a submerged land, covered by many fathoms of water, for the shallows of the river of my illustration, and some powerful ocean current, such as the gulf or arctic stream, for the river itself, and we at once arrive at a consistent theory of the boulder-clay and its origin. Nor must we deem it a thing improbable, that a country like Scotland, which lies between the fifty-fifth and the fifty-ninth degrees of north latitude, should be visited every year by icebergs. Newfoundland lies from five to eight degrees to the south of Scotland, and yet its northern shores are included in that vast cake of ice which, when winter sets fairly in, is found to stretch continuously, though in a winding line, over the surface of the ocean, from Nova Zembla in the Old World to Labrador in the New; and the drift ice-floes in spring, borne southwards on the Arctic current, brush every season over its southern shores, or ground by hundreds upon its great bank; nor do they finally disappear until they reach the fortieth, and, in at least one recorded instance, the thirty-sixth, degree of north latitude. I need scarce remind you that the temperature of a country depends on other causes than its distance from the equator or the pole. The isothermal line, or line of mean temperature, of the capital of Iceland, *Reikiavik*, in latitude 64°, is nearly as high as that of St. John's, the capital of Newfoundland, in latitude 47°; and *old* York, in the fifty-fourth degree of north latitude, enjoys as much average warmth throughout the year as *New* York, in the forty-first degree. Now, the causes which give to countries in the same latitudes climates so strangely different are known not to be permanent causes : temperature is found to depend on the disposition of land and sea, and the position, not of the geographical pole, which is single and centrical in each hemisphere, but of the pole of greatest cold, which, in at least the northern hemisphere, is double, and not centrical,—Asia having one, and America another; and if, as is generally held, there be a correspondence amounting almost to identity between the poles of greatest cold and the magnetic poles, then these poles are not fixed, but oscillating. Nor are we left to infer on merely general grounds that the climate of our country *may* have been at one time greatly more severe than it is now. There is also *zoological* evidence that it *was* greatly more severe. It is a curious and significant fact, that the group of shells found in the boulder-clay, resting over the scratched and grooved rocks, and accompanying the scratched and polished pebbles, is essentially a boreal or semi-arctic group. This little shell from the boulder-clay of Caithness,—the *Trophon scalariformis* or *Fusus scalariformis*, which, from its small size, seems to have escaped the fate that crushed its larger contemporaries into fragments, is not now found living on our coasts, though it still exists in considerable abundance in the seas of Greenland ; and several of its neighbours in the clay, such as *Tellina proxima* and *Astarte Borealis*, are of the same northern character. Nay, in cases in which the shells of the boulder-clay still live in our seas, we find those of a northern character, such as the *Cyprina Islandica*, that, though not rare on the shores of Scotland, is vastly more abundant on those of Iceland, occurring, not in the present British, but in the present Icelandic proportions. The *Cyprina Islandica* is one of the most common shells of the clay, and, as its name testifies, one of the most common shells of Iceland ; but it is by no means one of the most common shells at the present time of our Scottish coasts.

The shells of the boulder-clay correspond in the group, not to the present shells of Scotland, but to the present shells of Iceland and the Northern Cape.

Further, we are not left merely to infer that icebergs *could* or *might* have grooved and worn down the rocks of the country: we learn from Sir Charles Lyell,—unquestionably a competent observer,—that he caught icebergs almost in the very fact of grooving and wearing down similar rocks. In his first work of *Travels through the United States*, he describes a visit which he paid to the coast of Nova Scotia, near Cape Blomidon :—'As I was strolling along the beach,' he says, 'at the base of a line of basaltic cliffs, which rise over ledges of soft sandstone, I stopped short at the sight of an unexpected phenomenon. The solitary inhabitant of a desert island could scarcely have been more startled by a human footprint in the sand than I was on beholding some recent furrows on a ledge of sandstone under my feet, the exact counterpart of those grooves of ancient date which I have so often attributed to glacial action. . . . On a recently-formed ledge I saw several straight furrows half an inch broad, some of them very nearly parallel, others slightly divergent; and, after walking about a quarter of a mile, I found another set of similar furrows, having the same general direction within about five degrees; and I made up my mind that, if these grooves could not be referred to the modern instrumentality of ice, it would throw no small doubt on the glacial hypothesis. When I asked my guide, a peasant of the neighbourhood, whether he had ever seen much ice on the spot where we stood, the heat was so excessive (for we were in the latitude of the south of France, 45 degrees north), that I seemed to be putting a strange question. He replied, that in the preceding winter [that of 1841] he had seen the ice, in spite of the tide, which ran at the rate of ten miles an hour, extending in one uninterrupted mass from the shore where we stood, to the opposite coast of Parrsborough, and that the ice-blocks, heaped on each other and frozen together, or packed at the foot of Cape Blomidon, were often fifteen feet thick, and were pushed along, when the tide rose, over the sandstone ledges. He also stated that fragments of the black stone which fell from the summit of the cliff,—a pile of which lay at its base,—were often frozen into the ice, and moved along with it. And I have no doubt that the hardness of these gravers, firmly fixed in masses of ice, which, though only fifteen feet thick, are often of considerable horizontal extent, has furnished sufficient pressure and mechanical power to groove the ledges of soft sandstone.'

Thus far Sir Charles. The boulder-clay is found in Scotland from deep beneath the sea level, where it forms the anchoring ground of some of our finest harbours, to the height of from six to nine hundred feet along our hill-sides. The travelled boulders to which it owes its name have been found as high as fourteen hundred feet. Up to the highest of these heights icebergs at one time operated upon our Scottish rocks. Scotland, therefore, must in that icy age have been submerged to the highest of these heights. It must have existed as three groups of islands,—the Cheviot, or southern group; the Grampian, or middle group; and the Ben Wyvis, or northern group.

Let me next advert to a peculiarity in the direction of the icebergs which went careering at this period over the submerged land. As shown by the lines and furrows which they have graven upon the rocks, their general course, with a few occasional divergences,—effects, apparently, of the line of the greater valleys,—was from west to east. It is further a fact, exactly correspondent in the evidence which it bears, that the trap eminences of the country,—eminences of hard rock rising amid districts of soft sandstone, or still softer shale,—have generally attached to their eastern sides sloping *tali* of the yielding strata out of which they rise, and which have been washed away from all their other sides. Every larger stone in a water-course, after the torrent fed by a thunder-shower has just subsided, shows, on the same principle, its trail of sand and shingle piled up behind it,—sand and shingle which it kept from being swept away; and the simple effect, when it occurs on the large scale, is known to the geologist as the phenomenon of 'Crag and Tail.' The rock upon which Edinburgh Castle stands, existing as the '*crag*,' and the sloping ridge which extends from the castle's outer moat to Holyrood, existing as the *tail*, may be cited as a familiar instance. We find the same phenomenon repeated in the Calton Hill, and in various other eminences in the neighbourhood; as also in the Castle Hill of Stirling. And in all these, and many other cases, the *tail* which the *crag* protected is turned towards the east, indicating that the current which in the lapse of ages scooped out the valleys at the sides of the protecting crags, and in many instances formed, by its eddies, hollows in advance of them, just as we find hollows in advance of the larger stones of the water-course of my illustration, was a current which flowed from the west. The testimony of the ice-grooved rocks, and of the eminences composed of *crag* and *tail*, bear, we see, in the same line.

Now, this westerly direction of the current seems to be exactly that which, reasoning from the permanent phenomena of nature, might be premised. There must have been trade winds in every period of the world's history, in which the earth revolved from west to east on its axis; and with trade winds the accompanying drift current. And, of consequence, ever since the existence of a great western continent, stretching far from south to north, there must have been also a gulf stream. The waters heaped up against the coasts of this western continent at the equator by the drift current ever flowing westwards, must have been always, as now, returning eastwards in the temperate zone, to preserve the general level of the ocean's surface. Ever, too, since winter took its place among the seasons, there must have been an arctic current. The ice and snows of the higher latitudes, that accumulated during the winter, must have again melted in spring and early summer; and a current must in consequence have set in as the seasons of these came on, just as we now see such a current setting in in these seasons in both hemispheres, which bears the ice of the antarctic circle far towards the north, and the ice of the arctic circle far towards the south. The point at which, in the existing state of things, the gulf stream and the arctic current come in contact is that occupied by the great bank of Newfoundland; and by some the very existence of the bank has been attributed to their junction, and to the vast accumulation of gravel and stone cast down year after year from the drift ice to the bottom, where these two great tides meet and jostle. Be this as it may, the number of boulders and the quantity of pebbles and gravel strewed over the bottom of the western portions of the Atlantic, in the line of the arctic current, from the confines of Baffin's Bay up to the forty-fifth degree of north latitude, must be altogether enormous. Captain Scoresby counted no fewer than five hundred icebergs setting out on their southern voyage on the arctic current at one time. And wherever there are shallows on which these vast masses catch the bottom, or grate over it,—shallows of from thirty to a hundred fathoms water, —we may safely premise that at the present time there is a boulder-clay in the course of formation, with a scratched and polished surface of rock lying beneath it, and containing numerous pebbles and boulders striated longitudinally. That the point where the gulf and arctic currents come in contact should now lie so far to the west, is a consequence of the present disposition of the arctic and western continents,— perhaps also of the present position of the magnetic pole. A different arrangement and position would give a different point of meeting; and it is as little improbable that they should have met in the remote past some two or three hundred miles to the *west* of what is now Scotland, as that in the existing period they should meet some two or three hundred miles to the *east* of what is now Newfoundland. The northern current would be deflected by the more powerful gulf stream into an easterly course, and would go sweeping over the submerged land in the direction indicated by the grooves and scratches, bearing with it, every spring, its many thousand gigantic icebergs, and its fields of sheet-ice many hundred square miles in extent. And these, armed beneath with great pebbles and boulders, or finding many such resting at the bottom, by grinding heavily along the buried surface,—like the rafts of my illustration along the bed of the river,—would gradually wear down the upper strata of the softer formations, leaving the clay which they had thus formed to be deposited over, and a little to the east of, the rocks that had produced it. It is further in accordance with this theory, that in Scotland generally, the deeper deposits of the boulder-clay occur on the eastern line of coast. The cutler, in whetting a tool with water on a flat Turkey stone, drives the grey milky dressings detached by the friction of the steel from the solid mass, to the end of the stone furthest from himself, and there they accumulate thick in the direction of the stroke. And so it is here. The rubbings of the great Scotch whetstone, acted upon by the innumerable gravers and chisels whetted upon it, and held down or steadied by the icebergs, have been carried in the easterly direction of the stroke, and deposited at the further, that is to say, the eastern, end of the stone.

But fearing I have already too much trespassed on your time and patience, I shall leave half told for the present the story of the Pleistocene period in Scotland. If, instead of presenting it to you as a piece of clear, condensed narrative, I have led you darkly to grope your way through it by a series of fatiguing inductions, you will, I trust, sustain my apology, when I remind you that this dreary ice-epoch in the history of our country still forms as debatable a *terra incognita* to the geologist as the dreary ice-tracts which surround the pole do to the geographer. We have been threading our twilight way through a difficult North-West Passage; and if our progress has been in some degree one of weariness and fatigue, we must remember that without weariness and fatigue no voyager ever yet explored

> 'The ice-locked secrets of that hoary deep
> Where fettered streams and frozen continents
> Lie dark and wild, beat with perpetual storm
> Of whirlwind and dire hail.'

'We might expect,' says Professor Sedgwick, 'that as we come close upon living nature, the characters of our old records would grow legible and clear. But just where we begin to enter on the history of the physical changes going on before our eyes, and in which we ourselves bear a part, our chronicle seems to fail us; a leaf has been torn out from Nature's book, and the succession of events is almost hidden from our eyes.' Now it is to this age of the drift-gravels and the boulder-clay that the accomplished Professor here refers as represented in the geologic record by a torn page; and though we may be disposed to view it rather as a darkened one,—much soiled, but certainly not wanting,—we must be content to bestow on its dim, half-obliterated characters, more time and care than suffice for the perusal of whole chapters in the earlier books of our history. And so, casting myself on your forbearance, I shall take up the unfinished story of the Pleistocene period in Scotland in my next address.

LECTURE SECOND.

Problem first propounded to the Author in a Quarry—The Quarry's Two Deposits, Old Red Sandstone and Boulder-Clay—The Boulder-Clay formed while the Land was subsiding—The Groovings and Polishings of the Rocks in the Lower Parts of the Country evidences of the fact—Sir Charles Lyell's Observations on the Canadian Lake District—Close of the Boulder-Clay Record in Scotland—Its Continuance in England into the Pliocene Ages—The Trees and Animals of the Pre-Glacial Periods—Elephants' Tusks found in Scotland and England regarded as the Remains of Giants—Legends concerning them—Marine Deposits beneath the Pre-Glacial Forests of England—Objections of Theologians to the Geological Theory of the Antiquity of the Earth and of the Human Race considered—Extent of the Glacial Period in Scotland—Evidences of Glacial Action in Glencoe, Gareloch, and the Highlands of Sutherland—Scenery of Scotland owes its Characteristics to Glacial Action—The Period of Elevation which succeeded the Period of Subsidence—Its Indications in Raised Beaches and Subsoils—How the Subsoils and Brick Clays were formed—Their Economic Importance—Boulder-Stones interesting Features in the Landscape—Their prevalence in Scotland—The more remarkable Ice-travelled Boulders described—Anecdotes of the 'Travelled Stone of Petty' and the Standing-Stone of Torboll—Elevation of the Land during the Post-Tertiary Period which succeeded the Period of the Boulder-Clay—The Alpine Plants of Scotland the Vegetable Aborigines of the Country—Panoramic View of the Pleistocene and Post-Tertiary Periods—Modern Science not adverse to the Development of the Imaginative Faculty.

I REMEMBER, as distinctly as if I had quitted it but yesterday, the quarry in which, some two-and-thirty years ago, I made my first acquaintance with a life of toil and restraint, and at the same time first broke ground as a geologist. It formed a section about thirty feet in height by eighty or a hundred in length, in the front of a furze-covered bank, a portion of the old coast line; and presented an under bar of a deep-red sandstone arranged in nearly horizontal strata, and an upper bar of a pale-red clay roughened by projecting pebbles and boulders. Both deposits at the time were almost equally unknown to the geologist. The deep-red sandstone beneath formed a portion of that ancient Old Red system which represents, as is now known, the second great period of vertebrate existence on our planet, and which has proved to the palæontologist so fertile a field of wonders : the pale clay above was a deposit of the boulder-clay, resting on a grooved and furrowed surface of rock, and containing in abundance its scratched and polished pebbles. Old Red Sandstone and boulder-clay ! a broad bar of each ;—such was the compound problem propounded to me by the Fate that dropped me in a quarry ; and I gave to both the patient study of years. But the older deposit soon became frank and communicative, and yielded up its organisms in abundance, which furnished me with many a curious little anecdote of their habits when living, and of the changes which had passed over them when dead ; and I was enabled, with little assistance from brother geologists, to give a history of the system to the world more than ten years ago. The boulder-clay, on the contrary, remained for years invincibly silent and sullen. I remember a time when, after passing a day under its barren *scaurs*, or hid in its precipitous ravines, I used to feel in the evening as if I had been travelling under the cloud of night, and had seen nothing. It was a morose and taciturn companion, and had no speculation in it. I might stand in front of its curved precipices, red, yellow, or grey (according to the prevailing colour of the rocks on which it rested), and might mark their water-rolled boulders of all kinds and sizes sticking out in bold relief from the surface, like the protuberances that roughen the rustic basements of the architect ; but I had no '*Open, Sesame*' to form vistas through them into the recesses of the past. And even now, when I have, I think, begun to understand the boulder-clay a little, and it has become sociable enough to indulge me with occasional glimpses of its early history in the old glacial period,—glimpses of a half-submerged land, and an iceberg-mottled sea, turbid with the comminuted débris of the rocks below, you will see how very much I have had to borrow from the labours of others, and that in worming my way into its secret, there are obscure recesses within its precincts into which I have failed to penetrate. Let us now, however, resume its half-told story.

There are appearances which lead us to conclude, that during the formation and deposition of the boulder-clay, what is now Scotland was undergoing a gradual subsidence,—gradually foundering amid the waves, if I may so speak, like a slowly-sinking vessel, and presenting, as century succeeded century, hills of lower and yet lower altitude, and an ever lessening area. I was gratified to find, that when reasoning out the matter for myself, and arriving at this conclusion from the examination of one special set of data, Mr. Charles Darwin was arriving at the same conclusion from the consideration of a second and entirely different set ; and Sir Charles Lyell,—from whom, on the publication of my views in the *Witness* newspaper some four years since, I received a kind and interesting note on the subject,—had also arrived at the same conclusion—North America being the scene of his observations—from the consideration of yet a third and equally distinct set. And in the *Geological Journal* for the present year, I find Mr. Joshua Trimmer and Mr. Austin arriving, from evidence equally independent, at a similar finding. We have all come to infer, in short, that previous to the Drift period the land had stood at a comparatively high level,—perhaps higher than it does now ; that ages of depression came on, during which the land sank many hundred feet, and the sea rose high on the hill-sides ; and that during these ages of depression the boulder-clay was formed. Let me state briefly some of the considerations on which we found.

The boulder-clay, I thus reasoned with myself, is generally found to overlie more deeply the lower parts of the country than those higher parts which approach its upper limit ; and yet the rocks on which it rests, in some localities to the depth of a hundred feet at even the level of the sea, bear as decidedly their groovings and polishings as those on which, eight hundred feet over the sea level, it reposes to but the depth of a yard or two. Now, had a rising land been subjected piecemeal to the grinding action of the icebergs, this would not have been the case. The higher rocks first subjected to their action would of course bear the groovings and furrowings; but the argillaceous dressings detached from them in the process, mixed with the stones and pebbles which the ice had brought along with it, would necessarily come to be deposited in the form of boulder-clay on the lower rocks; and ere these lower rocks could be brought, by the elevation of the land, within reach of the grinding action of the icebergs, they would be so completely covered up and shielded by the deposit, that the bergs would fail to come in contact with them. They would go sweeping, not over the rocks themselves, but over the clay by which the rocks had been covered up; and so we may safely infer that, had the boulder-clay been formed during an elevating period, the lower rocks, where thickly covered by the clay, would not be scratched and grooved as we now find them, or, where scratched and grooved, would not be thickly covered by the clay. The existing phenomena, deep grooves and polished striæ, on rocks overlaid at the present sea-level to a great depth by the boulder-clay, demand for their production the reverse condition of a sinking land, in which the lower rocks are first subjected to the action of the icebergs, and the higher rocks after them. The quarrier, when he has to operate on some stratum of rock on a hill-side, has to commence his labours below, and to throw the rubbish which he forms behind him, leaving ever an open *face* in front; for, were he to reverse the process, and commence *above*, the accumulating débris, ever seeking downwards, would at length so choke up the working as to arrest his labours. And such, we infer from the work done, must have been the course of operations imposed by the conditions of a sinking land on the icebergs of the glacial period: they began their special course of action at the hill-foot, and operated upon its surface upwards as the sea arose. Again, Mr. Darwin's reasonings were mainly founded on the significant fact, that in numerous instances travelled boulders of the ice period may be found on levels considerably higher than those of the rocks from which they were originally torn. And though cases of transport from a lower to a higher level could and would take place during a period of subsidence, when the sea was rising or the land sinking, it is impossible that it could have taken place during an elevating period, when the sea was sinking or the land rising.[1] A flowing sea, to use a simple illustration, frequently carries shells, pebbles, and sea-weed from the level of ebb to the level of flood;—it brings them from a low to a high level: whereas an ebbing sea can but reverse the process, by bringing them from a high level to a low.

For the facts and reasonings of Sir Charles Lyell on the subject, I must refer you,—as they are incapable of being abridged without being injured—to that portion of his first work of Travels in America which treats of the Canadian Lake District. But the following are his conclusions:— '*First*,' he says, 'the country acquired its present geographical configuration, so far as relates to the older rocks, under the joint influence of elevating and denuding operations. *Secondly*, a gradual submergence then took place, bringing down each part of the land successively to the level of the waters, and then to a moderate depth below them. Large islands and bergs of floating ice came from the north, which, as they grounded on the coast and on shoals, pushed along all loose materials of sand and pebbles, broke off all angular and projecting points of rock, and, when fragments of hard

[1] See Mr. Trimmer's last paper on Boulder-Clays, *Journal of the Geological Society*, May 1858, p. 171.—W. S.

stone were frozen into their lower surfaces, scooped out grooves in the subjacent solid strata. *Thirdly*, after the surface of the rocks had been smoothed and grated upon by the passage of innumerable icebergs, the clay, gravel, and sand of the Drift were deposited; and occasionally fragments of rock, both large and small, which had been frozen into glaciers, or taken up by coast-ice, were dropped here and there at random over the bottom of the ocean, wherever they happened to be detached from the melting ice. *Finally*, the period of re-elevation arrived, or of that intermittent upward movement in which the old coast lines were excavated and the ancient sand bars or osars laid down.' Such are the conclusions at which Sir Charles Lyell arrived a few years since respecting the Canadian Lake District; and he states, in the note to which I have referred, that he has ever since been applying them to Scotland. Our country, during the chill and dreary period of the boulder-clay, seems to have been settling down into the waves, like the vessel of some hapless Arctic explorer struck by the ice in middle ocean, and sinking by inches amid a wild scene of wintry desolation.

There are a few detached localities in Scotland where the remains of beds of stratified sand and gravel have been detected underlying the boulder-clay; and in some of these in the valley of the Clyde, Mr. Smith of Jordanhill found on a late occasion shells of the same semi-arctic character as those which occur in the clay itself. And with these stratified beds the record in Scotland closes; whereas in England we find it carried interestingly onward from the Pleistocene period, first into the newer, and then into the older, Pliocene ages. I stated incidentally in my former address, that some of the mosses of the sister kingdom, unlike those of our own country, are older than the Drift period; and, from the existence of these under the Drift gravels and brown clay, it has been inferred by Mr. Trimmer, that as the trees which enter into their composition grew upon the surface of what is now England, where they now lie, previous to the period of the boulder-clay, and as the boulder-clay is, as shown by its remains, decidedly marine, it must have been deposited during a period of depression, when what had been a forest-bearing surface was lowered beneath the level of the sea. None of the trees of these ancient pre-glacial forests seem to be of extinct species: the birch and Scotch fir are among their commonest forms, especially the fir. I find it stated, however, as a curious fact, that along with these, the *Abies Excelsa*, or Norwegian spruce-pine, is found to occur,—a tree which, though introduced by man into our country, and now not very rare in our woods, has not been of *indigenous* growth in any British forest since the times of the boulder-clay. Though the species continued to live in Norway, it became extinct in Britain; and it has been suggested, that as it was during the Drift period that it disappeared, it may have owed its extirpation to the depression of the land, while its contemporaries the birch and fir were preserved on our northern heights. When this Norwegian pine flourished in Britain, the island was inhabited by a group of quadrupeds now never seen associated, save perhaps in a menagerie. Mixed with the remains of animals still native to our country, such as the otter, the badger, and the red deer, there have been found skeletons of the *Lagomy*, or tail-less hare, now an inhabitant of the cold heights of Siberia, and horns of the rein-deer, a species now restricted in Europe to Northern Scandinavia, and those inhospitable tracts of western Russia that border on the Arctic Sea. And with these boreal forms there were associated, as shown by their bones and tusks, the elephant, the rhinoceros, and the hippopotamus, all, however, of extinct species, and fitted for living under widely different climatal conditions from those essential to the well-being of their intertropical congeners.[1] Scotland,

[1] The true mammoth, with the tichorine rhinoceros and the musk buf-

though it has proved much less rich than England in the remains of the early Pleistocene mammals, has furnished a few well-attested elephantine fossils. In the summer of 1821, in the course of cutting the Union Canal, there was found in the boulder-clay near Falkirk, on the Clifton Hall property, about twenty feet from the surface, a large portion of the tusk of an elephant, three feet three inches in length and thirteen inches in circumference; and such was its state of keeping when first laid open, that it was sold to an ivory-turner by the labourers that found it, and was not rescued from his hands until a portion of it had been cut up for chessmen. Two other elephants' tusks were found early in 1817 at Kilmaurs[1] in Ayrshire, on a property of the Earl of Eglinton,—one of them so sorely decayed that it could not be removed; but a portion of the other, with the rescued portion of the Falkirk tusks, may be seen in the Museum of our Edinburgh University, which also contains, I may here mention, the horn of a rhinoceros, found at the bottom of a morass in Forfarshire, but which, in all probability, as it stands alone among the organisms of our mosses, had been washed out of some previously formed deposit of the Drift period. Scotland seems to have furnished several other specimens of elephantine remains; but as they were brought to light in ages in which comparative anatomy was unknown, and men believed that the human race had been of vast strength and stature in the primeval ages, but were fast sinking into dwarfs, they were regarded as the remains of giants. Some of the legends to which the

falo, are the leading types of the mammalian fauna of the Glacial Drift epoch. The remains of *hippopotamus* would be washed out of older beds.—W. S.

[1] At a later period (December 1829), similar elephantine tusks were found thirty-four feet beneath the surface, in boulder-clay overlying the quarry of Greenhill, also in Kilmaurs parish; and they may now be seen in the Hunterian Museum, Glasgow.

bones of these supposed giants served to give rise in England occupy a place in the first chapter of the country's history, as told by the monkish chroniclers, and have their grotesque but widely-known memorials in Gog and Magog, the wooden giants of Guildhall: our Scottish legends of the same class are less famous; but to one of their number, —charged with an argument in behalf of the temperance cause of which our friends the teetotallers have not yet availed themselves,—I may be permitted briefly to refer, in the words of one of our elder historians. 'In Murray land,' says the believing Hector Boece, ' is the Kirke of Pette, quhare the bones of Litell Johne remainis in gret admiration of pepill. He hes bene fourtene feet of hicht, with squaire membres effering thairto. Six yeirs afore the coming of this work to licht (1520) we saw his henche bane, as meikle as the haill banes of ane manne; for we schot our arme into the mouthe thairof; be quhilk appeirs how strang and squaire pepill greu in oure regeoun afore thay were effeminat with lust and intemperance of mouthe.'

Under these pre-glacial forests of England there rests a marine deposit, rich in shells and quadrupedal remains, known as the Norwich or Mammaliferous Crag; and beneath it, in turn, lie the Red and Coralline Crags—members of the Pliocene period. In the Mammaliferous Crag there appear a few extinct shells, blent with shells still common on our coasts. In the Red Crag the number of extinct species greatly increases, rising, it is now estimated, to thirty per cent. of the whole; while in the Coralline Crag the increase is greater still, the extinct shells averaging about forty per cent.[1] In these deposits some of our best-known molluscs appear in creation for the first time. The common edible oyster (*Ostrea edulis*) occurs in

[1] The known species of shells in the Coralline Crag amount to three hundred and forty. Of these, seventy-three are living British species. See Woodward's *Manual*, part iii. p. 421.—W.S.

the Coralline Crag, but in no older formation, and with it the great pecten (*Pecten maximus*), the horse mussel (*Modiola vulgaris*), and the common whelk (*Buccinum undatum*). Other equally well-known shells make their advent at a still later period; the common mussel (*Mytilus edulis*), the common periwinkle (*Littorina littorea*), and, in Britain at least, the dog-whelk (*Purpura lapillus*), first appear in the overlying Red Crag, and are not known in the older Coralline formation. By a certain very extended period, represented by the Coralline Crag, the edible oyster seems to be older than the edible mussel, and the common whelk than the common periwinkle; and I call your special attention to the fact, as representative of a numerous *class* of geological facts that bear on certain questions of a semi-theological character, occasionally mooted in the religious periodicals of the day. There are few theologians worthy of the name who now hold that the deductions of the geologists regarding the earth's antiquity are at variance with the statements of Scripture respecting its first creation, and subsequent preparation for man. But some of them do seem to hold that the scheme of reconciliation, found sufficient when this fact of the earth's antiquity was almost the only one with which we had to grapple, should be deemed sufficient still, when science, in its onward progress, has called on us to deal with this new fact of the very unequal antiquity of the plants and animals still contemporary with man, and with the further fact, that not a few of them must have been living upon the earth thousands of years ere he himself was ushered upon it,—facts of course wholly incompatible with any scheme of interpretation that would fix the date of their first appearance only a few *natural* days in advance of that of his own. We have no good reason to hold that the human species existed upon earth during the times of the boulder-clay: such a belief would conflict, as shown by the antiquity of the ancient and existing coast lines, with our received chronologies of the race. But long previous to these times, the Norwegian spruce pine and the Scotch fir were natives of the pre-glacial forests of our country; at even an earlier period the common periwinkle and edible mussel lived in the seas of the Red Crag deposits; and at a still earlier time, the great pecten, the whelk, and the oyster, in those of the Coralline Crag. We can now no more hold, as geologists, that the plants and animals of the existing creation came into being only a few hours or a few days previous to man, than that the world itself came into being only six thousand years ago; and we do think we have reason to complain of theologians who, ignorant of the facts with which we have to deal, and in no way solicitous to acquaint themselves with them, set themselves coolly to criticise our well-meant endeavours to reconcile the Scripture narrative of creation with the more recent findings of our science, and who pronounce them inadmissible, not because they do not effect the desired reconciliation, but simply because they are new to theology. They should remember that the *difficulty* also is new to theology; that enigmas cannot be solved until they are first propounded; that if the riddle be in reality a new one, the answer to it must of necessity be new likewise; and as this special riddle has been submitted to the geologists when the theologians were unaware of its existence, it must not be held a legitimate objection, that geologists, who feel that they possess, as responsible men, a stake in the question, should be the first to attempt solving it. If, however, it be, as I suspect, with our facts, not with our schemes of reconciliation, that the quarrel in reality lies,—if it be, in particular, with the special fact of the unequal antiquity of the existing plants and animals, and the comparatively recent introduction of man,—I would fain urge the objectors to examine ere they decide, and not rashly and in ignorance to commit themselves against truths which every day must render more palpable and clear, and which are destined long to outlive all cavil and opposition.

With respect to the antiquity of our race, we have, as I have said, no good grounds to believe that man existed upon the earth during what in Britain, and that portion of the Continent which lies under the same lines of latitude, were the times of the boulder-clay and Drift gravels. None of the human remains yet found seem more ancient than the historic period, in at least the older nations: it is now held that the famous skeletons of Guadaloupe belonged to men and women who must have lived since the discovery of America by Columbus; and if in other parts of the world there have been detected fragments of the human frame associated with those of the long extinct animals, there is always reason to conclude that they owe such proximity to that burying propensity to which I have already adverted, or to accidents resulting from it, and not to any imaginary circumstance of contemporarity of existence. If man buries his dead in the Gault or the London Clay, human remains will of course be found mingled with those of the Gault or the London Clay; but the evidence furnished by any such mixture will merely serve to show, not that the existences to which the remains belonged had lived in the same age, but simply that they had been deposited in the same formation. Nor can I attach much value to the supposed historic records of countries such as Egypt, in which dynasties are represented as having flourished thousands of years ere the era of Abraham. The chronicles of all nations have their fabulous introductory portions. No one now attaches any value to the record of the eighty kings that are said to have reigned in Scotland between the times of Fergus the First and Constantine the Bold; or to that portion of old English history which treats of the dynasty of Brutus the Parricide, or his wars with the giants. All the ancient histories have, as Buchanan tells us, in disposing of the English claims, their beginnings obscured by fable; nor is it probable that the Egyptian history is an exception to all the others, or that its laboriously inscribed and painfully interpreted hieroglyphics were more exclusively devoted to the recording of real events than characters simpler of form and easier of perusal. If, as some contend, man has been a denizen of this world for some ten or twelve thousand years, what, I would ask, was he doing during the first five or six thousand? It was held by Sir Isaac Newton, that the species must have been of recent introduction on earth, seeing that all the great human discoveries and inventions, such as letters, the principles of geometry and arithmetic, printing, and the mariner's compass, lie within the historic period. The mind of man could not, he inferred, have been very long at work, or, from its very constitution, it would have discovered and invented earlier; and all history and all archæological research bear out the inference of the philosopher. The older civilized nations lie all around the original centre of the race in Western Asia; nor do we find any trace of a great city older than Nineveh, or of a great kingdom that preceded in its rise that of Egypt. The average life of great nations does not exceed twelve, or at most fifteen, hundred years; and the first great nations were, we find, living within the memory of letters. Geology, too, scarce less certainly than Revelation itself, testifies that the last-born of creation was man, and that his appearance on earth is one of the most recent events of which it submits the memorials to its votaries.

But to return: The glacial or ice period in Scotland seems to have extended from the times of the stratified beds, charged with sub-arctic shells, which underlie the boulder-clay, until the land, its long period of depression over, was again rising, and had attained to an elevation less by only fifty or a hundred feet than that which it at present maintains. Such is the height over the sea level, of the raised beach at Gamrie in Banffshire; and in it the arctic shells last appear. And to the greatly-extended sub-arctic period in Scotland there belong a class of appearances which have been adduced in support of a *glacial* as opposed to an *iceberg* theory. But there is in reality no antagonism in the case. After examining not a few of our Highland glens, especially those on the north-western coast of the country, I have arrived at the conviction, that Scotland had at one time its glaciers, which, like those of Iceland, descended along its valleys, from its inland heights, to the sea. And as in most cases certain well-marked accompaniments of the true glacier, such as those lateral and transverse moraines of detached rock and gravel that accumulate along their sides and at their lower terminations, are wanting in Scotland, it is inferred that great currents must have swept over the country since the period of their existence, and either washed their moraines away, or so altered their character and appearance that they can be no longer recognised *as* moraines. Of course, this sweeping process might have taken place during that period of profound subsidence when the boulder-clay was formed, and in a posterior period of more partial subsidence, which is held to have taken place at a later time and under milder climatal conditions, and which is said to have brought down the land to its present level from a considerably higher one. In many localities there rests over the true boulder-clay an argillaceous or gravelly deposit, in which the masses and fragments of rock are usually angular, and which, even where the boulder-clay is shell-bearing, contains no shells. There are other localities in which a similar deposit also *underlies* the boulder-clay; and these deposits, upper and lower, are in all probability the débris of glaciers that existed in our country during the ice-era,—the lower deposit being the débris of glaciers that had existed previous to the glacial period of subsidence, and the upper that of glaciers which had existed posterior to it, and when the land was rising. The evidence is, I think, conclusive, that glaciers there were. I examined, during the autumn of last year, the famous Glencoe, and can now entertain no more doubt that a glacier once descended along the bottom of that deep and rugged valley, filling it up from side to side to the depth of from a hundred and fifty to two hundred feet, than that an actual glacier descends at the present day along the valley of the Aar or of the Grindelwald. The higher precipices of Glencoe are among the most rugged in the kingdom : we reach a certain level; and, though no change takes place in the quality of the rock, all becomes rounded and smooth, through the agency, evidently, of the vanished ice-river, whose old line of surface we can still point out from the continuous mark on the sides of the precipices, beneath which all is smooth, and above which all is rugged, and whose scratchings and groovings we can trace on the hard porphyry descending towards the Atlantic, even beyond where the sea occupies the bottom of the valley. The lines and grooves running in a reverse direction to those of the icebergs, for their course is towards the west, are distinctly discernible as far down as Ballachulish ferry. Similar marks of a great glacier in the valley of the Gareloch have been carefully traced and shrewdly interpreted by Mr. Charles M'Laren. But nowhere have I seen the evidence of glacial action more decided than in the Highlands of Sutherland, over which I travelled last August more than a hundred and fifty miles, for the purpose of observation. There is scarce a valley in that wild region, whether it open towards the northern or western Atlantic, or upon the German Ocean, that in this ungenial period was not cumbered, like the valleys of the upper Alps, by its burden of slowly-descending ice. Save where, in a few localities on the lower slopes of the hills, the true boulder-clay appears, almost all the subsoil of the country, where it *has* a subsoil, is composed of a loose, unproductive glacial débris; almost every prominence on the mountain-sides is rounded by the long protracted action of the ice; and in many instances the surfaces of the rocks bear the characteristic groovings and scratchings as distinctly as if it had performed its work upon them but yesterday. Let me, however, repeat the remark, that the iceberg and glacial theories, so far from being antagonistic, ought rather to be regarded as equally indispensable parts of one and the same theory, —parts which, when separated, leave a vast amount of residual phenomena to puzzle and perplex, that we find fully accounted for by their conjunction. And why not conjoin them? The fact that more than four thousand square miles of the interior of Iceland are covered by glaciers, is in no degree invalidated by the kindred fact that its shores are visited every spring by hundreds of thousands of icebergs.

The glaciers of Scotland have, like its icebergs, contributed their distinctive quota to the scenery of the country. The smoothed and rounded prominences of the hills, bare and grey amid the scanty heath, and that often after a sudden shower gleam bright to the sun, like the sides and bows of windward-beating vessels wet by the spray of a summer gale, form well-marked features in the landscapes of the north-western parts of Sutherland and Ross, especially in the gneiss and quartz-rock districts. The lesser islets, too, of these tracts, whether they rise in some solitary lochan among the hills, or in some arm of the sea that deeply indents the coast, still bear the rounded form originally communicated by the ice, and in some instances remind the traveller of huge whales heaving their smooth backs over the brine. Further, we not unfrequently see the general outline of the mountains affected;—all their peaks and precipices curved backwards in the direction *whence* the glacier descended, and more angular and abrupt in the direction *towards which* it descended. But it is in those groups of miniature hills, composed of glacial débris, which so frequently throng the openings of our Highland valleys, and which Burns so graphically describes in a single line as

'Hillocks dropt in Nature's careless haste,'

that perhaps the most pleasing remains of our ancient glaciers are to be found. They seem to be modified moraines, and usually affect regular forms, resembling in some instances the roofs of houses, and in some the bottoms of upturned ships; and, grouped thick together, and when umbrageous with the graceful birch, or waving from top to base with the light fronds of the lady-fern and the bracken, they often compose scenes of a soft and yet wild loveliness, from which the landscape gardener might be content to borrow, and which seem to have impressed in a very early age the Celtic imagination. They constitute the fairy Tomhans of Highland mythology; and many a curious legend still survives, to tell of benighted travellers who, on one certain night of the year, of ghostly celebrity, have seen open doors in their green sides, whence gleams of dazzling light fell on the thick foliage beyond, and have heard voices of merriment and music resounding from within; or who, mayhap, incautiously entering, have listened entranced to the song, or stood witnessing the dance, until, returning to the open air, they have found that in what seemed a brief half-hour half a lifetime had passed away. There are few of the remoter valleys of the Highlands that have not their groups of fairy Tomhans,—memorials of the age of ice.

After the lapse of ages,—but who can declare their number?—the period of subsidence represented by the boulder-clay came to a close, and a period of elevation succeeded. The land began to rise; and there is a considerable extent of superficial deposits in Scotland which we owe to this period of elevation. It is the main object of the ingenious work of Mr. Robert Chambers on Raised

Beaches to show that there were pauses in the elevating process, during which the lines against which the waves beat were hollowed into rectilinear terraces, much broken, it is true, and widely separated in their parts, but that wonderfully correspond in height over extensive areas. It is of course to be expected, that the higher and more ancient the beach or terrace, the more must it be worn down by the action of the elements, especially by the descent of watercourses; and as the supposed beaches intermediate between the strongly-marked ancient coast line which I have already described at such length, and certain upper lines traceable in the moorland districts of the country, occur in an agricultural region, the obliterating wear of the plough has been added to that of the climate. After, however, all fair allowances have been made, there remain great difficulties in the way. I have been puzzled, for instance, by the fact that Scotland presents us with but two lines of water-worn caves,—that of the present coast line, and that of the old line immediately above it. Mr. Chambers enumerates no fewer than fifteen coast lines intermediate between the old coast line and a coast line about three hundred feet over it; and in the range of granitic rocks which skirt on both sides the entrance of the Cromarty Firth, there are precipices fully a hundred yards in height, and broadly exposed to the stormy north-east, whose bases bear their double lines of deeply-hollowed caverns. But they exhibit no third, or fourth, or fifth line of caves. Equally impressible throughout their entire extent of front, and with their enclosed masses of chloritic schist and their lines of fault as thickly set in their brows as in their bases, they yet present no upper storeys of caves. Had the sea stood at the fifteen intermediate lines for periods at all equal in duration to those in which it has stood at the ancient or at the existing coast line, the taller precipices of the Cromarty Sutors would present their seventeen storeys of excavations; and excavations in a hard granitic gneiss. that varied from twenty to a hundred feet in depth would form marks at least as indelible as parallel roads on the mountain sides, or mounds of gravel and débris overtopping inland plains, or rising over the course of rivers. The want of lines of caves higher than those of the ancient coast line would seem to indicate, that though the sea may have remained long enough at the various upper levels to leave its mark on soft impressible materials, it did not remain long enough to excavate into caverns the solid rocks.

But though the rise of the land may have been comparatively rapid, there was quite time enough during the term of upheaval for a series of processes that have given considerable variety to the subsoils of our country. Had the land been elevated at one stride, almost the only subsoil of what we recognise as the agricultural region of Scotland would have been the boulder-clay, here and there curiously inlaid with irregular patches of sand and gravel, which occur occasionally throughout its entire thickness, and which were probably deposited in the forming mass by icebergs, laden at the bottom with the sand and stones of some sea-beach, on which they had lain frozen until floated off, with their burdens, by the tide. But there elapsed time enough during the upheaval of the land, to bring its boulder-clay deposits piecemeal under the action of the tides and waves; and hence, apparently, the origin of not a few of our lighter subsoils. Wherever the waves act at the present time upon a front of clay, we see a separation of its parts taking place. Its finer argillaceous particles are floated off to sea, to be deposited in the outer depths; its arenaceous particles settle into sand-beds a little adown the beach; its pebbles and boulders form a surface stratum of stones and gravel, extending from the base of the *scaur* to where the surf breaks at the half-tide line. We may see a similar process of separation going on in ravines of the boulder-clay swept by a streamlet. After every shower the stream comes down brown and turbid with the more argillaceous portions of the deposit; accumulations of sand are swept to the gorge of the ravine, or cast down in ripple-marked patches in its deeper pools; beds of pebbles and gravel are heaped up in every inflection of its banks; and boulders are laid bare along its sides. Now, a separation by a sort of washing process of an analogous character seems to have taken place in the materials of the more exposed portions of the boulder-clay, during the emergence of the land; and hence, apparently, those extensive beds of sand and gravel which in so many parts of the kingdom exist in relation to the clay as a superior or upper subsoil; hence, too, occasional beds of a purer clay than that beneath, divested of a considerable portion of its arenaceous components, and of almost all its pebbles and boulders. This *washed* clay,—a re-formation of the boulder deposit,—cast down mostly in insulated beds in quiet localities, where the absence of currents suffered the purer particles, held in suspension by the water, to settle, forms, in Scotland at least,—with, of course, the exception of the ancient fire-clays of the Coal Measures, —the true brick and tile clays of the agriculturist and architect. There are extensive beds of this washed clay within a short distance of Edinburgh; and you might find it no uninteresting employment to compare them, in a leisure hour, with the very dissimilar boulder-clays over which they rest. Unlike the latter, they are finely laminated: in the brick-beds of Portobello I have seen thin streaks of coal-dust, and occasionally of sand, occurring between the layers; but it is rare indeed to find in them a single pebble. They are the washings, in all likelihood, of those boulder-clays which rise high on the northern flanks of the Pentlands, and occur in the long flat valley along which the Edinburgh and Glasgow Railway runs,—washings detached by the waves when the land was rising, and which, carried towards the east by the westward current, were quietly deposited in the lee of Arthur Seat and the neighbouring eminences,—at that time a small group of islands. The only shells I ever detected in the brick-clay of Scotland occurred in a deposit in the neighbourhood of St. Andrews, of apparently the same age as the beds at Portobello.[1] They were in a bad state of keeping; but I succeeded in identifying one of the number as a deep-sea Balanus, still thrown ashore in considerable quantity among the rocks to the south of St. Andrews. In this St. Andrews deposit, too, I found the most modern nodules I have yet seen in Scotland, for they had evidently been hardened into stone during the recent period; but, though I laid them open by scores, I failed to detect in them anything organic. Similar nodules of the Drift period, not unfrequent in Canada and the United States, are remarkable for occasionally containing the only ichthyolite found by Agassiz among seventeen hundred species, which still continues to live, and that can be exhibited, in consequence, in duplicate specimens,—the one fit for the table in the character of a palatable viand,—the other for the shelves of a geological museum in the character of a curious ichthyolite. It is the *Mallotus villosus*, or Capelen (for such is its market-name), a little fish of the arctic and semi-arctic seas. 'The *Mallotus* is abundant,' says Mr. James Wilson, in his admirable *Treatise on Fishes*, 'in the arctic seas, where it is taken in immense profusion when approaching the coasts to spawn, and it is used as the principal bait for cod. A few are cured and brought to this country in barrels, where they are sold, and used as a relish by the curious in wines.'

Let me next call your attention to the importance, in an economic point of view, of the great geologic events which gave to our country its subsoils, more especially the boulder-clay. This deposit varies in value, according to the nature

[1] See Note at the end of the Lectures.

of the rocks out of which it was formed; but it is, even where least fertile, a better subsoil than the rock itself would have been; and in many a district it furnishes our heaviest wheat soils. To the sand and gravel formed out of it, and spread partially over it, we owe a class of soils generally light, but kindly; and the brick clays are not only of considerable value in themselves, but of such excellence as a subsoil, that the land which overlies them in the neighbourhood of Edinburgh still lets at from four to five pounds per acre. I suspect that, in order to be fully able to estimate the value of a subsoil, one would need to remove to those rocky lands of the south that seem doomed to hopeless barrenness for want of one. It is but a tedious process through which the minute lichen or dwarfish moss, settling on a surface of naked stone, forms, in the course of ages, a soil for plants of greater bulk and a higher order; and had Scotland been left to the exclusive operation of this slow agent, it would be still a rocky desert, with perhaps here and there a strip of alluvial meadow by the side of a stream, and here and there an insulated patch of mossy soil among the hollows of the crags; but, though it might possess its few gardens for the spade, it would have no fields for the plough. We owe our arable land to that geologic agent which, grinding down, as in a mill, the upper layers of the surface rocks of the kingdom, and then spreading over the eroded strata their own débris, formed the general basis in which the first vegetation took root, and in the course of years composed the vegetable mould. A foundering land under a severe sky, beaten by tempests and lashed by tides, with glaciers half choking up its cheerless valleys, and with countless icebergs brushing its coasts and grating over its shallows, would have seemed a melancholy and hopeless object to human eye, had there been human eyes to look upon it at the time; and yet such seem to have been the circumstances in which our country was placed by Him who, to 'perform his wonders,'

> 'Plants his footsteps in the sea,
> And rides upon the storm,'

in order that at the appointed period it might, according to the poet, be a land

> 'Made blithe by plough and harrow.'

From the boulder-clay there is a natural transition to the boulders themselves, from which the deposit derives its name. These remarkable travelled stones seem, from the old traditions connected with some of them, to have awakened attention and excited wonder at an early period, long ere Geology was known as a science; nor are they without their share of picturesqueness in certain situations. You will perhaps remember how frequently, and with what variety of aspect, Bewick, the greatest of wood-engravers, used to introduce them into the backgrounds of his vignettes. 'A rural scene is never perfect,' says Shenstone, a poet of no very large calibre, but the greatest of landscape gardeners, 'without the addition of some kind of building: I have, however, known,' he adds, 'a scaur of rock in great measure supplying the deficiency.' And the justice of the poet's canon may be often seen exemplified in those more recluse districts of the country which border on the Highlands, and where a huge rock-like boulder, roughened by mosses and lichens, may be seen giving animation and cheerfulness to the wild solitude of a deep forest glade, or to some bosky inflection of bank waving with birch and hazel on the side of some lonely tarn or haunted streamlet. Even on a dark sterile moor, where the pale lichen springs up among the stunted heath, and the hairy club-moss goes creeping among the stones, some vast boulder, rising grey amid the waste, gives to the fatigued eye a reposing point, on which it can rest for a time, and then let itself out on the expanse around. Boulder-stones are still very abundant in Scotland, though for the last century they have been gradually disappearing from the more cultivated tracts where there were fences or farm steadings to be built, or where they obstructed the course of the plough. We find them occurring in every conceivable situation : high on hill-sides, where the shepherd crouches beside them for shelter in a shower; deep in the open sea, where they entangle the nets of the fisherman on his fishing banks; on inland moors, where in some remote age they were laboriously rolled together to form the Druidical circle or Pict's House; or on the margin of the coast, where they had been piled over one another at a later time, as protecting bulwarks against the waves. They are no longer to be seen in this neighbourhood in what we may term the agricultural region; but they still occur in great numbers along the coast, within the belt that intervenes between high and low water, and on an upper moorland zone over which the plough has not yet passed. Mr. Charles M'Laren describes, in his admirable little work on *The Geology of Fife and the Lothians*, a boulder of mica schist weighing from eight to ten tons, which rests, among many others, on one of the Pentland Hills, and which derives an interest from the fact that, as shown by the quality of the rock, the nearest point from which it could have come is at least fifty miles away. A well-known greenstone boulder of still larger size may be seen at the line of half-ebb, about half-way between Leith and Portobello. But though about ten feet in height, it is a small stone, compared with others of its class both in this country and the Continent. The *rock*, as it is well termed (for it is a mass of granite weighing fifteen hundred tons), on which the colossal statue of Peter the Great at St. Petersburg is placed, is a travelled boulder, which was found dissociated from every other stone of its kind in the middle of a morass; and Sir Roderick Murchison describes, in one of his papers on the Northern Drift, a Scandinavian boulder thirty feet in height by one hundred and forty in circumference. Most, if not all the boulders which we find in this part of the country on the lower zone have been washed out of the boulder-clay. Wherever we find a group of boulders on the portion of sea-bottom uncovered by the ebb, we have but to look at the line where the surf breaks when the sea is at full, and there we find the clay itself, with its half-uncovered boulders projecting from its yielding sides, apparently as freshly grooved and scratched as if the transporting iceberg had been at work upon them but yesterday.

I must again adduce the evidence of Sir Charles Lyell, to show that masses of this character *are* frequently ice-borne. 'In the river St. Lawrence,' we find him stating in his *Elements*, 'the loose ice accumulates on the shoals during the winter, at which season the water is low. The separate fragments of ice are readily frozen together in a climate where the temperature is sometimes thirty degrees below zero, and boulders become entangled with them; so that in spring, when the river rises on the melting of the snow, the ice is floated off, frequently conveying the boulders to great distances. A single block of granite fifteen feet long by ten feet both in breadth and height, and which could not contain less than fifteen hundred cubic feet of stone, was in this way moved down the river several hundred yards, during the late survey in 1837. Heavy anchors of ships lying on the shore have in like manner been closed in and removed. In October 1806 wooden stakes were driven several feet into the ground at one part of the banks of the St. Lawrence at high-water mark, and over them were piled many boulders as large as the united force of six men could roll. The year after, all the boulders had disappeared, and others had arrived, and the stakes had been drawn out and carried away by the ice.'

Our Scottish boulders,—though in many instances immediately associated, as in this neighbourhood, with the boulder-clay, and in many others, as in our moorland districts, with the bare rock,—occur in some cases associated with the superficial sands and gravels, and rest upon or over these. And in these last instances they must have been the subjects of a course of ice-borne voyagings subsequent to the earlier course, and when the land was rising. Even during the last sixty years, though our winters are now far from severe, there have been instances in Scotland of the transport of huge stones by the agency of ice; and to two of these, as of a character suited to throw some light on the boulder voyagings of the remote past, I must be permitted to refer.

Some of my audience may have heard of a boulder well known on both sides of the Moray Firth as the 'Travelled Stone of Petty,'—a district which includes the Moor of Culloden, and at whose parish church Hector Boece saw the gigantic bones of the colossal Little John. The *Clach dhu n-Aban*, or black stone of the white bog,—for such is the graphically descriptive Gaelic name of the moss,—measures about six feet in height by from six to seven feet in breadth and thickness, and served, up to the 19th of February 1799, as a *march-stone* between the properties of Castle Stuart and Culloden. It lay just within flood-mark, near where a little stream empties itself into a shallow sandy bay. There had been a severe, long-continued frost throughout the early part of the month; and the upper portions of the bay had acquired, mainly through the agency of the streamlet, a continuous covering of ice, that had attained, round the base of the stone, which it clasped fast, a thickness of eighteen inches. On the night of the 19th the tide rose unusually high on the beach, and there broke out a violent hurricane from the east-south-east, accompanied by a snow storm. There is a meal mill in the immediate neighbourhood of the stone; and when the old miller,—as he related the story to the late Sir Thomas Dick Lauder,—got up on the morning of the 20th, so violent was the storm, and so huge the snow-wreaths that blocked up every window and door, and rose over the eaves, that he could hardly make his way to his barns,—a journey of but a few yards; and in returning again from them to his dwelling, he narrowly escaped losing himself in the drift. In looking towards the bay, in one of the pauses of the storm, he could scarce credit his eyesight; the immense *Clach dhu n-Aban* had disappeared,—vanished,— gone clean off the ground; and he called to his wife in astonishment and alarm, that the 'meikle stane was awa.' The honest woman looked out, and then rubbed her eyes, as if to verify their evidence; but the fact was unquestionable,—the 'meikle stane' certainly 'was awa;' and there remained but a hollow pit in the sand, with a long shallow furrow, stretching from the pit outwards to where the snow rhime closed thick over the sea, to mark where it had been. When, however, the weather cleared up, the stone again became visible, lying out in the sands uncovered by the ebb, seven hundred and eighty feet from its former position. In the evening of the day, the neighbours flocked out by scores, to examine the scene of so extraordinary a prodigy. Where the stone had lain they found but the deep dent, connected by the furrow which lay athwart the bay in the line of the hurricane with the stone itself, around the base of which there still projected a thick cornice of ice. In its new position the stone still lies; and only a few years ago,—mayhap still,—a wooden post which marked the point where the two contiguous properties met, marked also the spot from which, after a rest of ages, it had set out on its short voyage.

My other case of boulder travelling,—in some respects a more curious case than the one related,—occurred early in the present century on the eastern coast of Sutherlandshire. Near the small hamlet of Torboll, in the upper part of Loch Fleet, there stood, about fifty years ago, a rude obelisk of undressed stone, generally regarded as Danish, which, though more ancient than authentic history, or even tradition, in the district, was less so than the old coast line, as it had been evidently erected, subsequent to the last change of level, on the flat marginal strip which intervenes between the old line and the sea. It rose in the middle of a swampy hollow, which protracted rains sometimes converted into a strip of water, and which was sometimes swept by the overflowings of the neighbouring river. On the eve of the incident which proved the terminating one in its history, the hollow, previously filled with rain-water, had been frozen to the bottom by a continued frost, which was, however, on the eve of breaking up ; and a dense fog lay thick in the valley, when a benighted Highlander, returning tipsy from a market by the light of the moon, came staggering in the direction of the standing stone, and in a drunken frolic set his bonnet on the top of it ; and then, wandering off into the mist, he lost sight of both stone and bonnet, and, failing to regain them, he had to return bareheaded to his home. The thaw came on ; the river rose over its banks ; the ice-cake around the obelisk floated high above the level, wrenching up the obelisk along with it, as the ice of the St. Lawrence wrenched up the stakes described by Sir Charles Lyell ; and both ice-cake and obelisk floated down the loch to the sea. As the morning broke,—a fierce morning of flood and tempest,—they were seen passing what some forty years ago was known as the Little Ferry ; and the alarm went abroad along the shores on both sides, that there was a *man* standing in the middle of the Loch on the floating ice, and in course of being swept out to the ocean. Poor man ! he had been crossing the river, it was inferred, when the ice broke up ; and though the enterprise was a somewhat perilous one, for the ice fragments were rushing furiously along on the wild tides of the loch, maddened by the inundation, a boat, double manned, shot out from the shore to the rescue, and soon neared the drifting ice-floe. It was ultimately seen, however, that the supposed man was but the Danish obelisk, bearing on its head a mysterious bonnet; and bonnet and obelisk were left to find their way to the German Ocean, in which it is probable they now both lie. These modern instances of boulder travelling may serve to show how huge stones originally associated with the boulder-clay may have come to rest on the arenaceous or gravelly deposits which overlie it. Through the second voyage of the Petty boulder, it was deposited on a recently formed bed of sand; and the standing-stone of Torboll may now rest on sea-shells that were living half a century ago.

It is held by geologists of high standing, that after the period of submergence represented by the boulder-clays of our country, the British islands were elevated to such a height over the sea-level, that their distinctive character as islands was lost, and the area which they occupy united to the main land in the character of a western prolongation of the great European continent. It was at this period, says Professor Edward Forbes, that Britain and Ireland received, over the upraised bed of the German Ocean, their Germanic flora,— the last acquired of the five floras which compose their vegetation. The evidence on the point, however, still seems somewhat meagre. I can have no doubt that the land stood considerably higher during this Post-Tertiary period than it does now. As shown by the dressed surfaces and rounded forms of many of the smaller islets of the north-western coasts of Scotland, and the markings at the bottom of its lochs and estuaries, and on the rocks along their shores, the latter glaciers must have descended from the central hills of the country far below the present sea-level; and we find some of the transverse moraines which they ploughed up before them in their descent existing as gravelly spits, that rise amid the waves, in the middle of long firths or at the entrance of deep bays. I have seen, too, on rocky coasts, considerably below the tide-line at flood, a sort of recent breccia formed by calcareous springs, which, as the stalagmitical matter could not have been deposited in places exposed to the diurnal washings of the sea, indicated a higher level of the land than now, at the time of its formation; and the submerged mosses of both Britain and Ireland,—mosses now existing in many localities far below the fall of the tide,—where not more ancient than the boulder-clay, bear evidence in the same line. But on this obscure passage in the geological history of our country I am unable, from at least actual observation, to say aught more: my few facts lie in the direction of Professor Forbes's theory, but they accompany it only a short way. There is a wide gap still unfilled. I may be permitted to remind you, that it is held by the Professor,—one of the most accomplished of our geologists,—that of the five British floras, we have two in Scotland,—the Germanic flora, and the semi-arctic or Scandinavian flora; that these were introduced into the country at different periods; and that while the Germanic flora dates from the times of the Post-Tertiary elevation of the land, the more ancient of the two—the semi-arctic or Scandinavian—dates from the preceding times of the boulder-clay. Nor does it appear in any degree more improbable that we should have the descendants of the plants of even the remoter period still vital on our hill-tops, than that we should have the descendants of some of its animals still living in our seas. It seems at first a curious problem, difficult of solution, that widely separated mountain summits should possess the same alpine plants,—that the summits of Ben Wyvis and Ben Lomond, for instance, or of Ben Nevis and Ben Muich Dhui, should have their species in common, while not a trace of them appears on the lower elevations between. But it simplifies the case to conceive of these alpine plants as the vegetable aborigines of the country, compelled by climatal invasion to shelter in its last bleak retreats, where the winter snows linger unwasted till midsummer, and the breeze is always laden with the chills of the old glacial period. They compose the *Celtic* portion of the Scottish flora, cooped up in their mountain recesses by the encroachments of those Germanic races of the plant family that flourish, in the altered atmosphere, on the more genial plains of the country, or on the sunny slopes of its lower hills. That language of flowers in which the ladies of Mohammedan countries have learned to converse is not unappropriately employed in giving expression to the various modes of a passion scarce less evanescent than the flowers themselves. But is it not passing strange, that we of Scotland should be called on to recognise in the transitory flowers of our sheltered low-lying plains and valleys, and of our high bleak moors and exposed mountain summits, the records of an antiquity so remote, that the stories told by the half-effaced hieroglyphics of Nineveh and of Egypt are of yesterday in comparison?

Here the exhibition of our facts illustrative of the Pleistocene and Post-Tertiary periods in Scotland properly ends. The existing evidence has been taken, though, of course, briefly and imperfectly, the extent and multiplicity of the subject considered; and, the record closed, a formal summary of the conclusions founded upon it should now terminate our history. Permit me, however, to present you, in conclusion, not with the formal summary, but a somewhat extended picture, of the whole, exhibited, panorama-like, as a series of scenes. The fine passage in the Autumn of Thomson, in which the poet lays all Scotland at once upon the canvas, and surveys it at a glance, must be familiar to you all :—

> ' Here awhile the Muse,
> High hovering o'er the broad cerulean scene,
> Sees Caledonia in romantic view ;
> Her airy mountains, from the waving main,

> Invested with a keen diffusive sky,
> Breathing the soul acute; her forests huge,
> Incult, robust, and tall, by Nature's hand
> Planted of old; her azure lakes between,
> Pour'd out extensive, and of watery wealth
> Full; winding deep and green, her fertile vales,
> With many a cool, translucent, brimming flood
> Wash'd lovely, from the Tweed (pure parent stream,
> Whose pastoral banks first heard my Doric reed,
> With, sylvan Jed, thy tributary brook),
> To where the North's inflated tempest foams
> O'er Orcas or Betubium's highest peak.'

Let us in like manner attempt calling up the features of our country in one continuous landscape, as they appeared at the commencement of the glacial period, just as the paroxysm of depression had come on, and bold headland and steep iron-bound islet had begun slowly to settle into the sea.

The general outline is that of Scotland, though harsher and more rugged than now, for it lacks the softening integument of the subsoils. Yonder are the Grampians, and yonder the Cheviots, and, deeply indenting the shores, yonder are the well-known estuaries and bays,—the firths of Forth, Tay, and Moray, and the long withdrawing lakes, Loch Katrine, and Loch Awe, and Loch Maree, and the far-gleaming waters of the deep Caledonian Valley, the Ness, and the Oich, and the Lochy. But though the summer sun looks down upon the scene, the snow-line descends beneath the top of even our second-class mountains; and the tall beetling Ben Nevis, and graceful Ben Lomond, and the broad-based Ben Muich Dhui, glitter in the sunshine, in their coats of dazzling white, from their summits half-way down to their bases. There are extended forests of the native fir on the lower plains, mingled with the slimmer forms and more richly-tinted foliage of the spruce pine. On the upper grounds, thickets of stunted willows and straggling belts of diminutive birches skirt the ravines and watercourses, and yellow mosses and grey lichens form the staple covering of the humbler hill-sides and the moors. But the distinctive feature of the country is its glaciers. Fed by the perpetual snows of the upper heights, the deeper valleys among the mountains have their rigid ice-rivers, that in the narrower firths and lochs of the western and northern coasts shoot far out, mole-like, into the tide. And, lo! along the shores, in sounds and bays never yet ploughed by the keel of voyager, vast groups of icebergs, that gleam white to the sun, like the sails of distant fleets, lie moveless in the calm, or drift slowly along in rippling tideways. Nor is the land without its inhabitants, though man has not yet appeared. The colossal elephant, not naked and dingy of coat, like his congener of the tropics, but shaggy, with long red hair, browses among the woods. There is a strong-limbed rhinoceros wallowing in yonder swamp, and a herd of rein-deer cropping the moss high on the hill-side beyond. The morse is basking on that half-tide skerry; and a wolf, swept seawards by the current, howls loud in terror from yonder drifting ice-floe. We have looked abroad on our future country in the period of the first local glaciers, ere the submergence of the land.

Ages pass, and usher in the succeeding period of the boulder-clay. The prospect, no longer that of a continuous land, presents us with a wintry archipelago of islands, broken into three groups by two deep ocean-sounds,—the ocean-sound of the great Caledonian Valley, and that of the broader but shallower valley which stretches across the island from the Clyde to the Forth. We stand full in front of one of these vast ocean-rivers,—the southern one. There are snow-enwrapped islets on either side. Can yonder thickly-set cluster be the half-submerged Pentlands? and yonder pair of islets, connected by a low flat neck, the eastern and western Lomonds? and yonder half-tide rock, blackened with algæ, and around which a shoal of porpoises are gambolling, the summit of Arthur Seat? The wide sound, now a rich agricultural valley, is here studded by its fleets of tall icebergs,—there cumbered by its level fields of drift-ice. Nature sports wantonly amid every variety of form; and the motion of the great floating masses, cast into shapes with which we associate moveless solidity, adds to the magical effect of the scene. Here a flat-roofed temple, surrounded by colonnades of hoar and wasted columns, comes drifting past; there a cathedral, furnished with towers and spire, strikes heavily against the rocky bottom, many fathoms beneath, and its nodding pinnacles stoop at every blow. Yonder, already fast aground, there rests a ponderous castle, with its curtained towers, its arched gateway, and its multitudinous turrets, reflected on the calm surface beneath; and pyramids and obelisks, buttressed ramparts, and embrasured watch-towers, with shapes still more fantastic,—those of ships, and trees, and brute and human forms,—crowd the retiring vista beyond. There is a scarce less marked variety of colour. The intense white of the field-ice, thinly covered with snow, and glittering without shade in the declining sun, dazzles the eye. The taller icebergs gleam in hues of more softened radiance,—here of an emerald green, there of a sapphire blue, yonder of a paly marble grey; the light, polarized by a thousand cross reflections, sports amid the planes and facets, the fissures and pinnacles, in all the rainbow gorgeousness of the prismatic hues. And bright over all rise on the distant horizon the detached mountain-tops, now catching a flush of crimson and gold from the setting luminary. But the sun sinks, and the clouds gather, and the night comes on black with tempest; and the grounded masses, moved by the violence of the aroused winds, grate heavily along the bottom; and while the whole heavens are foul with sleet and snow-rack, and the driving masses clash in rude collision, till all beneath is one wide stunning roar, the tortured sea boils and dashes around them, turbid with the comminuted débris of the fretted rocks below.

The vision belongs to an early age of the boulder-clay: it changes to a later time; and the same sea spreads out as before, laden by what seem the same drifting ice-floes. But the lower hills, buried in the profound depths of ocean, are no longer visible; the Lammermuirs have disappeared; and the slopes of Braid and Duddingstone, with

> "North Berwick Law, with cone of green,
> And Bass amid the waters;"

and we can only determine their place by the huger icebergs that lie stranded and motionless on their peaks; while the lesser masses drift on to the east. Moons wax and wane, and tides rise and fall; and still the deep current of the gulf-stream flows ever from the west, traversing the wide Atlantic, like some vast river winding through an enormous extent of meadow; and, in eddying over the submerged land, it arranges behind the buried eminences, in its own easterly line, many a long trail of gravel and débris, to form the Crag and Tail phenomenon of future geologists. As we extend our view, we may mark, far in the west, where the arctic current, dotted white with its ice-mountains and floes, impinges on the gulf stream; and where, sinking from its chill density to a lower stratum of sea, it gives up its burden to the lighter and more tepid tide. A thick fog hangs over the junction, where the warmer waters of the west and south encounter the chill icy air of the north; and, steaming forth into the bleak atmosphere like a seething caldron, the cloud, when the west wind blows, fills with its thick grey reek the recesses of the half-foundered land, and obscures the prospect.

Anon there is another change in the dream.* The long period of submergence is past; the country is again rising; and, under a climate still ungenial and severe, the glaciers lengthen out seawards, as the land broadens and extends, till the northern and western Highlands seem manacled in ice. Even the lower hill-tops exhibit an alpine vegetation, beautiful, though somewhat meagre; while in the firths and bays, the remote ancestors of many of our existing shells that thrive in the higher latitudes, still mix, as at an earlier period, with shells whose living representatives are now to be sought on the coasts of northern Scandinavia and Greenland. Ages pass; the land rises slowly over the deep, terrace above terrace; the thermal line moves gradually to the north; the line of perpetual snow ascends beyond the mountain summits; the temperature increases; the ice disappears; the semi-arctic plants creep up the hill-sides, to be supplanted on the plains by the leafy denizens of happier climates; and at length, under skies such as now look down upon us, and on nearly the existing breadth of land, the human period begins. The half-naked hunter, armed with his hatchet or lance of stone, pursues the roe or the wild ox through woods that, though comparatively but of yesterday, already present appearances of a hoar antiquity; or, when the winter snows gather around his dwelling, does battle at its beleaguered threshold with the hungry wolf or the bear. The last great geologic change takes place; the coast line is suddenly elevated; and the country presents a new front to the sea. And on the widened platform, when yet other ages have come and gone, the historic period commences, and the light of a classical literature falls for the first time on the incidents of Scottish story, and on the bold features of Scottish character.

It is said that modern science is adverse to the exercise and development of the imaginative faculty. But is it really so? Are visions such as those in which we have been indulging less richly charged with that poetic pabulum on which fancy feeds and grows strong, than those ancient tales of enchantment and *faery* which beguiled of old, in solitary homesteads, the long winter nights. Because science flourishes, must poesy decline? The complaint serves but to betray the weakness of the class who urge it. True, in an age like the present,—considerably more scientific than poetical,—science substitutes for the smaller poetry of fiction, the great poetry of truth; and as there is a more general interest felt in new revelations of what God has wrought, than in exhibitions of what the humbler order of poets have half-borrowed, half-invented, the disappointed dreamers complain that the 'material laws' of science have pushed them from their place. As well might the Arab who prided himself upon the beauty of some white tent which he had reared in some green oasis of the desert, complain of the dull tools of Belzoni's labourers, when engaged in clearing from the sands the front of some august temple of the ancient time. It is not the tools, it might be well said to the complainer, that are competing with your neat little tent; it is the sublime edifice, hitherto covered up, which the tools are laying bare. Nor is it the material laws, we may, on the same principle, say to the poets of the querulous cast, that are overbearing your little inventions, and making them seem small; but those sublime works and wonderful actings of the Creator which they unveil, and bring into comparison with yours. But from His works and His actings have the masters of the lyre ever derived their choicest materials; and whenever a truly great poet arises,—one that will add a profound intellect to a powerful imagination,—he will find science not his enemy, but an obsequious caterer and a devoted friend. He will find sermons in stones, and more of the suggestive and the sublime in a few broken scaurs of clay, a few fragmentary shells, and a few green reaches of the old coast line, than versifiers of the ordinary calibre in their once fresh gems and flowers,—in sublime ocean, the broad earth, or the blue firmament and all its stars.

LECTURE THIRD.

The Poet Delta (Dr. Moir)—His Definition of Poetry—His Death—His Burial-place at Inveresk—Vision, Geological and Historical, of the Surrounding Country—What it is that imparts to Nature its Poetry—The Tertiary Formation in Scotland—In Geologic History all Ages contemporary—Amber the Resin of the *Pinus succinifer*—A Vegetable Production of the Middle Tertiary Ages—Its Properties and Uses—The Masses of Insects enclosed in it—The Structural Geology of Scotland—Its Trap Rock—The Scenery usually associated with the Trap Rock—How formed—The Cretaceous Period in Scotland—Its Productions—The Chalk Deposits—Death of Species dependent on Laws different from those which determine the Death of Individuals—The Two great Infinites.

THE members of the Philosophical Institution of Edinburgh enjoyed the privilege last season of listening to one of the sweetest and tenderest of modern British poets eloquently descanting on the history of modern British poetry. Rarely had master established for himself a better claim to teach. And, regarding the elegant volume produced on that occasion, so exquisite in its taste and so generous in its criticisms, it may justly be said that perhaps its *only*, at all events its gravest defect, is the inevitable one that, in exhibiting all that during the bypast generation was most characteristic and best in the poesy of our country, it should have taken no cognizance of the poetry of Delta. Dr. Moir had just finished his course, but his volume had not yet appeared, when, urged by a friend, I perhaps too rashly consented to contribute two lectures to a course then delivering in the native town of the poet; and in one of these I expressed the conviction to which I gave utterance last season in this place, that there is no incompatibility between the pursuit of geologic science and a genial development of the poetic faculty. Dr. Moir had honoured my address with his presence; he had listened with apparent attention to a view very much opposed, as I was told after the breaking up of the meeting, to one which he himself had promulgated to the Institution only a few weeks before; and on the publication of his little volume he politely sent me a copy, accompanied by a kind note, in which he referred to the point apparently at issue between us, as involving rather a seeming than a real difference. 'Our antagonism respecting the relations of poetry and science,' he said, 'is, I doubt not, much more apparent than real, and arises simply from the opposite aspects in which we have regarded the subject.' I read his work with interest; and at first deemed the difference somewhat more than merely apparent. I found the lecturer speaking of 'staggering blows' inflicted on the poetry of the age by science in not a few formidably prosaic shapes,—in the shape, among the rest, of 'geological exposition;' and of 'rocks stratified by the geologists as satins are measured by mercers,' and, in consequence, no longer redolent of that emotion of the sublime which was wont to breathe forth of old from broken crags and giddy precipices. But his *definition* of poetry re-assured me, and set all right again. 'Poetry,' he said, 'may be defined to be objects or subjects seen through the mirror of imagination, and descanted on in harmonious language; and if so, it must be admitted that the very exactness of knowledge is a barrier to the laying on of that colouring by which facts can be invested with the illusive hues of poetry. Wherever light penetrates the obscure and illuminates the uncertain, we may rest assured that a demesne has been lost to the realms of imagination.' Now, if such be poetry, I said, and such the conditions favourable to its development, the poets need be in no degree jealous of the geologists. The stony science, with buried creations for its domains, and half an eternity charged with its annals, possesses its realms of dim and shadowy fields, in which troops of fancies already walk like disembodied ghosts in the old fields of Elysium, and which bid fair to be quite dark and uncertain enough for all the purposes of poesy for centuries to come.

Alas! only a few weeks after, amid hundreds of his sorrowing friends and townsmen, I followed the honoured remains of the poet to the grave; and heard, in that old picturesque burying-ground which commands on its green ridge the effluence of the Esk, the shovelled earth falling heavy on the coffin-lid. It was a lovely day of chequered shadow and sunshine; and the wide firth slept silently in the calm, with a dream-like spectrum of the heavens mirrored on its bosom. From the sadness of the present my thoughts let themselves out upon the past. I stood among the groves on a grassy mound which had been reared by the old Roman invader greatly more than a thousand years before; and I bethought me how, on visiting the place a few twelvemonths previous, for the first time, I had first of all sought out the burying-ground of the family of the deceased,—a spot endeared to every lover of poesy by those tenderest and sweetest of 'domestic verses' which show how truly, according to Cowper, 'the poet's lyre' had been 'the poet's heart;' and how I had next set myself to trace, as next in interest, the remains of that stern old people whose thirst of conquest and dominion had led them so far. And lo! like a dream remembered in a dream, as the crowd broke up and retired, the visions of that quiet day were again conjured up before me, but bearing now a felt reference to the respected dead, and accompanied by the conviction that, had we been destined to meet, and to compare at length our respective views, we should have found them essentially the same.

On that rising ground, so rich in historic associations, both Somerset and Cromwell had planted their cannon, and it had witnessed the disaster at Pinkie, and the headlong flight of the dragoons of Cope. But, passing over the more recent scenes, the vision of a forest-covered country rose before me,—a vision of the ancient aboriginal woods rising dusky and brown in one vast thicket, from the windings of the Esk to the pale brow of the Pentlands. Nor was the landscape without its human figures. The grim legionaries of the Proconsul of Augustus were opening with busy axes a shady roadway through the midst; and the incessant strokes of the axe and the crash of falling trees echoed in the silence throughout the valley. And then there arose another and earlier vision, when the range of semicircular heights which rise above the ancient Saxon borough, with its squat tower and antique bridge, existed as the coast line, and the site of the town itself as a sandy bay, swum over by the sea-wolf and the seal; and the long ridge now occupied by garden and villa, church and burying-ground, as a steep, gravelly bar, heaped up in the vexed line, where the tides of the river on the one hand contended with the waves of the firth on the other; and the Esk, fed by the glaciers of the interior, whose blue gleam I could mark on the distant Lammermuirs and the steeper Pentlands, rolled downwards, a vast stream, that filled from side to side the ample banks which, even when heaviest in flood, it scarce half-fills now; while a scantier and dingier foliage than before, composed chiefly of taper spruce and dark pine, roughened the lower plains, and flung its multitudinous boughs athwart the turbid and troubled eddies. And then there arose yet other and remoter scenes. From a foreground of weltering sea I could mark a scattered archipelago of waste uninhabited islands, picturesquely roughened by wood and rock; and near where the Scottish capital now stands, a submarine volcano sent forth its slim column of mingled smoke and vapour into the sky. And then there rose in quick succession scenes of the old Carboniferous forests: long withdrawing lakes, fringed with dense thickets of the green Calamite, tall and straight as the masts of pinnaces, and inhabited by enormous fishes, that glittered through the transparent depths in their enamelled armour of proof; or glades of thickest verdure, where the tree-fern mingled its branch-like fronds with the hirsute arms of the gigantic club-moss, and where, amid strange forms of shrub and tree no longer known on earth, the stately Araucarian reared its proud head two hundred feet over the soil; or yet again, there rose a scene of coral bowers and encrinal thickets, that glimmered amid the deep green of the ancient ocean, and in which, as in the groves sung by Ovid, the plants were sentient, and the shrinking flowers bled when injured. And, last of all, on the further limits of organic life a thick fog came down upon the sea, and my excursions into the remote past terminated, like the voyage of an old fabulous navigator, in thick darkness. Each of the series of visions, whether of the comparatively recent or the remote past, in which I at that time indulged, had employed the same faculties and gratified the same feelings; and though, in surveying the stuff out of which they had been sublimed, I could easily say where the historic ended and the geologic began, no corresponding line indicated in the visions themselves where the poetry ended and the prose began. The visions, whether historic or geologic, 'were of imagination all compact.' They all involved the same processes of *mind*—though, of course, in this instance, *mind* of a humbler order and ruder texture—as those exhibited in the sweet and fragrant verse of the poet himself,—as those exercised, let me say, in his vision on 'Mary's Mount,' when, with quiet graves above, and surrounded by quiet fields, he saw the contending hosts of a former day thronging the lower ground, and,

> 'With hilt to hilt, and hand to hand,
> The children of our mother land
> To battle came;'

or when he called up, after the lapse of half a lifetime, how when, in a wintry morning, he had journeyed before daybreak, a happy boy, along the frozen Esk, and saw

> 'In the far west the Pentland's gloomy ridge
> Belting the pale blue sky, whereon a cloud,
> Fantastic, grey, and tinged with solemn light,
> Lay like a dreaming monster, and the moon,
> Waning, above its silvery rim upheld
> Her horns, as 'twere a spectre of the past.'

I shall continue to hold, therefore, that there was no real difference between the views of the poet and those which I myself entertain, but that, as he himself well expressed it, our 'apparent antagonism arose simply from the opposite aspects in which we had viewed the subject.' He had been thinking of but stiff diagrams and hard names,—of dead strata measured off, in 'geological exposition,' by the yard and the mile, and enveloped in the obscuring folds of a Babylonish phraseology: while I, looking through the crooked characters and uncouth sounds in which the meanings of the science are locked up, to the meanings themselves, was luxuriating among the strange wild narratives and richly poetic descriptions of which its pregnant records consist.

What is it, let me ask, that imparts to Nature its poetry? It is not in Nature itself; it resides not either in dead or organized matter,—in rock, or bird, or flower; 'the deep saith, It is not in me, and the sea saith, It is not in me.' It is in mind that it lives and breathes: external nature is but its storehouse of subjects and models; and it is not until these are called up as images, and invested with 'the light that never was on land or sea,' that they cease to be of the earth earthy, and form the ethereal stuff of which the visions of the poet are made. Nay, is it not mainly through that associative faculty to which the sights and sounds of present nature become suggestive of the images of a nature not present, but seen within the mind, that the landscape pleases, or that we find beauty in its woods or beside its streams, or the impressive and the sublime among its mountains and rocks? Nature is a vast tablet, inscribed with signs, each of which has its own significancy, and becomes poetry in the mind when read; and geology is simply the key by which myriads of these signs, hitherto undecipherable, can be unlocked and perused, and thus a new province added to the poetical domain. We are told by travellers, that the rocks of the wilderness of Sinai are lettered over with strange characters, inscribed during the forty years' wanderings of Israel. They testify, in their very existence, of a remote past, when the cloud-o'ershadowed tabernacle rose amid the tents of the desert; and who shall dare say whether to the scholar who could dive into their hidden meanings they might not be found charged with the very songs sung of old by Moses and by Miriam, when the sea rolled over the pride of Egypt? To the geologist every rock bears its inscription engraved in ancient hieroglyphic characters, that tell of the Creator's journeyings of old, of the laws which He gave, the tabernacles which He reared, and the marvels which He wrought, —of mute prophecies wrapped up in type and symbol,—of earth gulfs that opened, and of reptiles that flew,—of fiery plagues that devastated on the dry land, and of hosts more numerous than that of Pharaoh, that 'sank like lead in the mighty waters;' and, having in some degree mastered the occult meanings of these strange hieroglyphics, we must be permitted to refer, in asserting the poetry of our science, to the sublime revelations with which they are charged, and the vivid imagery which they conjure up. But our history lags in its progress, while we discuss the poetic capabilities of the study through which its records are read and its materials derived.

In the deposits of that Tertiary division of the geologic formation which represents in the history of the globe the period during which mammals began to be abundant, and in which the great Cuvier won his laurels, Scotland is one of the poorest of European countries. Save for the comparatively recent discovery of Tertiary beds in the island of Mull by a nobleman fitted by nature either to adorn the literature or extend the science of his country, the geological historian would have to pass direct from the Pleistocene beds, with their grooved and polished pebbles and their semi-arctic shells, to the Chalk fossils of Banff and Aberdeen. But the discovery of his Grace the Duke of Argyll furnishes us with an interesting glimpse of a middle period widely different in its character from either the Cretaceous system or the boulder-clay. In the island of Mull, in a headland that rises about 130 feet over the sea, there occur, interposed between thick beds of trap, three comparatively thin beds of a grey arenaceous shale, charged with fossil leaves, as beautifully spread out, and with their ribs and veins as distinctly visible, as if they had been preserved in the herbarium of a botanist. Most of them belong to extinct species of existing families of dicotyledonous trees, such as the plane and the buckthorn, mingled, however, with narrow linear leaves of cone-bearing trees, which are supposed to belong, in this instance, to a species of yew, and with what seem the fronds of fern and the stems of equisetaceæ. Some of the beds of coal which have been long known to occur among the traps of the island of Mull are regarded by the Duke of Argyll as prolongations of these Tertiary leaf-beds, so mineralized by some metamorphic action as to have lost the organic structure. There must have been vast accumulations of leaves ere they could have yielded beds of coal. The middle or second bed of the three his Grace describes as peculiarly rich in the leafy impressions of this ancient period; and I need scarce say how suggestive the glimpse is which is furnished us by these buried layers of the foliage of Tertiary forests in Scotland, of which no other known memorial remains. You all remember Coleridge's fine comparison of the sorely-worn sails of the vessel in which the ancient mariner performed his voyage of peril and prodigy, to

> 'Brown skeletons of leaves that lay
>   The forest brook alóng,
> When the ivy tod is heavy with snow,
> And the owlet whoops to the wolf below;'

and you must have often marked the extreme delicacy of those deposited leaves, macerated during the winter season at the bottom of some woodland pool, which suggested the poet's simile. In that Tertiary period to which the leaf-beds of Mull belong, it would seem that extensive forests, chiefly of deciduous trees, shed year after year their summer coverings of leaves, some of which fell, and some of which were blown by the autumnal gusts, into the streams of the country, and were swept down by the current to lakes or estuaries, where they lay gradually resolving into such brown skeletons as caught the eye of Coleridge. We learn further, that there were forces active at the time, of which at any later period we have had no examples in the British islands. One of the leaf-beds described by his Grace is overlaid by a bed of volcanic ashes or tuff seven feet thick; another by a bed of similar ashes mixed with chalk flints, twenty feet thick; and yet another—the topmost layer—bears over it a bed of overflowing columnar basalt, forty feet thick. The volcanic agencies were active in what is now Scotland during the ages of its Tertiary forests.

The only Tertiary fossils of Scotland yet discovered are these forest and fern leaves of the Mull deposits. Their place in the great geologic division to which they belong is still definitely to fix; but some of our higher geologists are, I find, disposed to refer them to the second Tertiary or Miocene epoch, though with considerable hesitation. They belong, it is probable, to a period not very widely removed from that of the richly fossiliferous Marlstone of Œningen, on the banks of the Rhine, with its vast abundance of plants, chiefly dicotyledonous,—of fishes specifically different from those which now exist, but of the existing genera,—of a fox, which only the comparative anatomist can distinguish from the recent species of this country,—and of reptiles generically akin to those of the United States. It is a curious fact that, both in its animal and vegetable productions, that part of the New World which borders upon the Atlantic in the temperate zone, from Carolina to the mouth of the St. Lawrence, still presents very much the appearance which was presented by the flora and fauna of Europe during the later Tertiary periods. It has been often remarked, in reference to human manners and the progress of civilisation, that all ages of the world may be regarded as contemporary. Man is still, in many of the South Sea Islands, what he was in our own country previous to the times of the Roman invasion; and there are provinces in Spain and Portugal in which neither the people nor the clergy have got beyond the semi-barbarism of the Middle Ages. Curiously enough, in geologic history also, though in a narrower and more restricted sense, all ages are contemporary. The Galapagos have their age of reptiles, New Zealand its age of birds, and New Holland its age of marsupial quadrupeds. These countries bear now, in not a few particulars, the character of the Oolitic period in our own country. Again, on the eastern coasts of North America we are presented with a vegetation greatly resembling that of some of the later Tertiary periods; and of several of its animals the type is still more ancient. America, though emphatically the *New* World in relation to its discovery by civilized man, is, at least in these regions, an *old* world in relation to geological type; and it is the so-called *Old* World that is in reality the *new* one. 'If we compare,' says Professor Agassiz, in his late admirable work, *Lake Superior*,—'if we compare a list of the fossil trees and shrubs from the Tertiary beds of Œningen with a catalogue of the trees and shrubs of Europe and North America, it will be seen that the differences scarcely go beyond those shown by the different floras of these continents under the same latitudes. But what is quite extraordinary and unexpected is the fact, that the European fossil plants of that locality resemble more closely the trees and shrubs which grow at present in the eastern parts of North America, than those of any other part of the world; thus allowing us to express correctly the difference between the opposite coasts of these continents, by saying that the present eastern American flora, and, I may add, the fauna also, have a more ancient character than those of Europe. The plants, especially the trees and shrubs growing in our days in the United States, are, as it were, old-fashioned; and the characteristic genera Lagoings, Chelydra, and the large Salamanders, with permanent gills, that remind us of the fossils of Œningen, are at least equally so: *they bear the marks of former ages.*' This interesting fact,—vouched for by assuredly no mean authority,—may enable us to conceive of the general aspect of our country, so far at least as its appearance depended on its vegetation, towards the close of the Miocene period. Old Scotland exhibited features in that age greatly resembling those presented to the Puritan Fathers by the forest-covered shores of New England little more than two centuries ago. But no family of man dwelt in its solitary woods; and, as shown by its widely spread deposits of trap-tuff, and its vast beds of overlying basalt, broken by faults and shifts, its ancient volcanoes had not yet died out, and it must have had its frequent earthquake-agues and shaking-fits.

There is, however, another witness besides the leaf-beds of the island of Mull, which we may properly call into court to give evidence regarding the Tertiary period in Scotland. It is known that from a very early time masses of amber have been occasionally furnished by the north-eastern shores of the kingdom, in especial by that extensive tract of coast which stretches from the Buchan-ness to the Firth of Tay; and the geologist now recognises amber as a vegetable production of the Middle Tertiary ages. It is the resin of an extinct pine, which the fossil botanist has only of late learned to term the *Pinus succinifer*, or amber pine, but which the Prussian peasantry, who gather amber on the southern shores of the Baltic, used for ages to associate with this substance, from its occurrence in a fossil state in the same beds as *amber wood*. The ornamental character of this precious resin seems to have been appreciated by the native Scotch at an early period: beads of amber have been found in the old sepulchral barrows of the kingdom. Its value, however, as we learn from the first notice of it which occurs in our written history,—that of Hector Boece,—has not been *always* appreciated. After describing it, not very inadequately, as 'ane maner of goum or electuar, hewit like gold, and sa attractive of natur, that it drawis stra, flax, or hemmes of claethis to it in the samen maner as does an adamant stone grow,' he goes on to say that 'twa year afore the comin af [his] buke to licht (1524) thair arrivit an gret lompe of this goum in Buchquhane, als meikle as an hens; and wes brocht hame by the herdes quhilk wer kepand thair bestis, to thair housis, and cassin in the fere. And becaus they fand an smell and odour thairwith, they scha to thair maister that it wes garand for the *in*sens that is maid in the kirkes. Thair maister wes ane rud man as thay wer; and tuk bot ane litell part thairof, and left the remanent part behind him as mater of litell effect. All the parts of this goum, quhen it wes broken, wes of hew of gold, and schone lyke the licht of an candell. The maist part of this goum or electuar wes destroyit be rud peple afore it cam to any wise mannis eirs; of quhome may be verifyet the proverb, "The sow cares not for balme." Als sone as I wes advertisit thairof, I maid sic diligence that ane pairt of it was brocht me at Aberdene.' I may add to this notice of the old chronicler, that up to a comparatively recent period, ornaments of amber, especially amber beads of large size, or, as they were termed by our ancestors, 'lamour beads,' were highly valued by the humbler Scotch. That mysterious attractive property which resided in this gem-like resin, and which has since been found pregnant with that wonderful science to which the substance has given its Greek name, *electrum*, threw a halo of mystery around it, that served to enhance its native beauty. The Laird of Dumbiedykes was, it must be confessed, neither a very fervent nor very poetical lover; but a lover he was; and yet he could find nothing more apt with which to compare the eyes of his mistress, when turned upon him in her gratitude, than to beads of amber. 'Dinna ye think,' said the laird, 'puir Jeanie's e'en, wi' the tears in them, glanced like lamour beads, Mr. Saddletree?'

To the geologist this precious gum of the Tertiary ages is fraught with a peculiar interest, from the circumstance that it forms the best of all matrices for the preservation of organisms of the more fragile kinds. Mosses, fungi, and liverworts, are plants of so delicate a structure, that they are rarely or never preserved in shale or stone; but specimens of all three have been found locked up in amber in a state of the most perfect keeping. And, besides containing fragments of the pine which produced it, it has been found to contain minute pieces of four other species of pine, with bits of cypresses, yews, junipers, oaks, poplars, beeches, etc.,— in all, forty-eight different species of shrubs and trees, which must have flourished in the forests where it grew, and which, 'viewed in the group, may be regarded as constituting,' says Professor Göppert, 'a flora of a North American character.' You will of course remark how directly this evidence bears on that of Professor Agassiz. The most remarkable organisms of the amber, are, however, its insects,—a kind of fossils suggestive of a very different poetry from that which Pope elaborated from them in his well-known simile :—

> ' Pretty in amber to observe the forms
> Of hairs, or straws, or dirt, or grubs, or worms :
> The things, we know, are neither rich nor rare,
> But wonder how the mischief they got there !'

Fossil insects occur in both the Secondary and Palæozoic divisions, but rarely indeed in a state of sufficient entireness to enable the entomologist to distinguish their species. Even in classing them into families and genera, our best writers on the subject, such as the Rev. Mr. Brodie, confess that some of the number are very imperfectly made out. In the amber, on the contrary, even the most delicate ephemeræ that ever sported for a single summer evening in a forest glade, and then perished as the night came on, are preserved in a state of perfect entireness. In the amber of Prussia eight hundred different kinds of insects have been determined, most of them belonging to species, and even genera, that appear to be distinct from any now known; while of the others, some are nearly related to indigenous species, and some seem identical with existing forms that inhabit the warmer climates of the south. From their great specific variety and abundance we may infer that insects then, as now, formed the most numerous division of the animal kingdom. Our entomologists reckon at the present time about eleven thousand species of recent British insects, —a number many times greater than that of all its other denizens of the animal kingdom united. You will scarce deem the riddle regarding the entombment of these fragile creatures in the amber, which so puzzled the poet, particularly a hard one : the process must have resembled that which we see going on in our pine-forests every summer. The little flutterers must have settled on the bleeding trunks of the *Pinus succinifer*, and stuck fast, and the after flow of the sap covered them over. They add an interesting feature, identical with that sung by the poet, to the odoriferous amber forests of the Tertiary. The hot sun is riding high over the recesses of one of these deep woods, never yet trodden by human foot, and lighting up the waved lines of delicate green with which spring, just passing into early summer, has befringed the dark pines, and the yet unwithered catkins of the poplar and plane, and the white blossoms of the buckthorn. The cave-bear and hyena repose in silence in their dens, and not a wandering breeze rustles among the young leafage.

> 'But hark! how through the peopled air
>   The busy murmur glows;
> The insect youth are on the wing,
> Eager to taste the honeyed spring,
> And float amid the liquid noon:
> Some lightly o'er the current skim,
> Some show their gaily gilded trim
>   Quick glaring to the sun.'

And lo! where the forest glade terminates in a brown primeval wilderness, the sunbeams fall with dazzling brightness on the trunk of a tall stately tree, just a little touched with decay; and it reflects the light far and wide, and gleams in strong contrast with the gloom of the bosky recesses beyond, like the pillar of fire in the wilderness relieved against the cloud of night. 'Tis a decaying pine of stateliest size, bleeding amber. The insects of the hour flutter around it; and when, beguiled by the grateful perfume, they touch its deceitful surface, they fare as the lords of creation did in a long posterior age, in that

> 'Serbonian bog,
> Betwixt Damiata and Mount Casius old,
> Where armies whole have sunk.'

But, as happened to so many of the heroes of classic history, death is fame here, and by dying they became immortal; for it is from the individuals who thus perish that future ages are yet to learn that the species which they represent ever existed, or to become acquainted with even the generic peculiarities by which they were distinguished.

The question still remains, Whence has the amber of our Scottish coasts been derived? It occurs *in situ* in Tertiary deposits in the neighbourhood of London : good specimens of considerable size have been found, for instance, in a clay-pit near Hyde Park corner, not a quarter of a mile from the site of the Crystal Palace. It occurs, too, in Prussia, in a clay-bed of considerable horizontal extent, of which the larger part lies under the waves of the Baltic, but which rises on some parts of the coast about forty feet over the level of that sea, and to which of late years a sort of classical interest has been given by a modern fiction, worthy, from its air of matter-of-fact truthfulness, of our own Defoe,—the *Amber Witch.* The black amber vein found by the pastor's little daughter is described in the story as occurring high in a wooded defile behind her father's parsonage, and as owing its black colour to the quantity of charcoal, *i.e.,* carbonized wood, which it contained. And in both particulars the description is true to the geology of the amber deposits. But we have no amber deposits in Scotland : had amber ever existed in connexion with the Tertiary beds of Mull, it would have shared, in all probability, from the close proximity of the trap, the fate of the great lumps of butter which that giant in the nursery story who used to eat knights and young ladies, employed in testing the heat of his oven ; and so we must look for its place, not on our shores, but in the seas by which they are washed. But it is here necessary that I should submit to you a brief outline of the structural geology of our country, not only that we may know in what direction to look for its Tertiary beds, but in order also that we may form such an acquaintance with the general framework of our subject, as it exists in space, as may guide us in all our after conceptions regarding it. Avoiding the prolixity of minute detail, I shall present you at present with but a few of the leading lines.

The great central nucleus of Scotland, presenting considerably more than fifteen thousand square miles of surface, consists of what we shall term, with the elder geologists, primary rocks,—granites, gneisses, mica-schists, quartz-rocks, and clay-slates. These extend in one direction from the southern base of the Grampians to the northern limits of Sutherlandshire, and from Peterhead and Aberdeen on the east to Glenelg and Loch Carron on the west. [Now, around this great primary mass there runs a ring of the sedimentary fossiliferous rocks, somewhat, though of course not with such unbroken regularity, as a frame runs round a picture, or as the metallic setting of a Cairngorm or pebble brooch surrounds the stone. Of these earlier fossiliferous rocks, known about the beginning of the present century as the Grauwacke, and now as the Silurians, the frame or ring contains but fragments,—a narrow strip along the flanks of the Grampians on the south, and a few detached patches along the shores of Banff on the north and east. But the ring or frame of the next oldest fossiliferous system, the Old Red Sandstone, is very nearly complete; and to such a breadth do we find it developed, especially in the southern and northern parts of the enclosing frame, that, with the addition of a few patches in the border counties of Scotland, we find it occupying nearly five thousand square miles of the surface of our country.[1]] Outside the Old

[1] The Old Red Sandstone *frame*, and its corresponding illustrations, no longer hold good. The geology of north-western Scotland has recently been investigated by Sir Roderick Murchison, from whose researches it appears that Silurian strata occupy a much wider area of that district than had been previously suspected. Aided by Mr. Peach's discovery of Lower Silurian fossils in the crystalline limestones of Sutherlandshire, Sir Roderick has succeeded in showing that from the Atlantic to the German Ocean there is a regular succession of strata in ascending order, representing the Laurentian gneiss of Canada and the

Red Sandstone frame there occurs to the south, in the line of the great flat valley which runs across the country from the Firth of Forth to that of the Clyde, a broad belt of the Coal Measures,—the system which succeeds to it in natural sequence; but on the east, west, and north, the Coal Measures and New Red Sandstone are wanting, and we find fragments of a ring of Lias, as at Applecross on the one coast, and at Cromarty and Shandwick on the other; and outside the Lias, considerable fragments of yet another and wider ring of the Oolite. The sea on the east coast, and both that and numerous outbursts of overlying trap on the west, covers up the ring which lies beyond; but the Chalk flints and Greensand fossils of Aberdeen and Banff shires on the one hand, and the Chalk flints of Mull and Caithness on the other, indicate its existence and its components. An outer ring or frame of Chalk and Greensand, more or less broken, surrounds on two, mayhap on three, sides, the central nucleus of the kingdom; and were the beds of the German and Atlantic Oceans to be laid dry to the depth of about fifty fathoms, and the area of Scotland to be proportionally extended, you would find formation succeeding formation, in crossing the ring from the nucleus outwards, as we find them succeeding each other in the south of England, when crossing the country from South Wales in the direction of London. Beyond this outer ring of Chalk there lie, it is more than probable, deposits of the Tertiary system. Of the Mull deposits on the west coast we at least know, though they occur in so disturbed and overflown a district, that they lie outside the Secondary deposits of the island; and again on the east coast, where the Tertiary deposits, which occupy so large a portion of

Cambrian and Lower Silurian rocks of Wales, and superposed upon these older formations in the great Old Red Sandstone of Caithness. See the abstract of Sir Roderick Murchison's paper in the Reports of the Leeds Meeting of the British Association.—G.

the south-eastern portion of England, outside the Chalk, lose themselves in the German Ocean, the dredge has found interesting trace of them far at sea running northwards, to form, apparently, our submarine belt or ring. It is stated by Woodward, in his *Geology of Norfolk*, that the oyster-fishers on that coast dredged up from a tract of oyster-beds near Happisburgh no fewer than two thousand grinders of mammoths in the course of thirteen years. Further, those parts of the Continent which lie opposite our eastern coasts, including Holland, Hanover, and the larger part of Denmark, all consist of deposits of the Tertiary system, which, trending westwards at a low angle, form, it is probable, no inconsiderable part of the bed of the German Ocean. Those beds, however, from which our Scottish amber is derived must lie deep in the sea, outside the Lias, the Oolite, the Greensand, and the Chalk; and our specimens are rare in consequence, because at great depths the bottom is little affected by tempests. Not less than eight hundred pounds weight of this substance has been thrown up on the coast of east Prussia by a single storm.

From the Tertiaries we would naturally pass, in our upward progress, to the Secondary deposits; and of these, the remains of the Cretaceous system, as exhibited in Banff and Aberdeen shires, would, of course, first solicit notice, as representative in Scotland of that portion of the Secondary period nearest our own,—the period with which this great middle division of the earth's history terminated. I must first, however, call your attention to a series of rocks which, without belonging to any of the three great sedimentary divisions, seem in our own country to have been contemporary with them all. I refer to the trap rocks of the kingdom. The Duke of Argyll found in the island of Mull, as has been already shown, thick beds of trap, tuffacious and basaltic, overlying beds of the Tertiary division. Again, in the Isle of Skye, Professor Edward

Forbes has detected trap beds which made their way to the surface, and overflowed the shells and corals of the Oolite, about the middle of the great Secondary period. 'The thick sheet of imperfectly columnar basalt,' says the Professor, 'which has so wide an extension in the island of Skye, and plays so important a part in the formation of the magnificent scenery of its coasts, was the product of a submarine eruption, which, if we regard the basalt as an overflow, has its geological date marked to a nicety, having occurred at the close of the middle and at the commencement of the upper Oolitic period.' Yet again, in the neighbourhood of Edinburgh, as well described by Mr. Charles M'Laren, there are traps of the Palæozoic division, —beds of stratified tuff, as among the rocks of the Calton Hill, for instance,—that belong to the early part of the Carboniferous period; and I have seen at Oban a conglomerate low in the Old Red Sandstone, formed chiefly of a trap, which even at that early time must have been a surface rock much exposed to denudation. We must regard, then, the trap rocks of Scotland as of all ages, from the earlier Palæozoic to the middle Tertiary periods. The great ganoidal fishes of the Devonian and Carboniferous ages, the huge reptiles of the Oolite, and the gigantic mammals of the Miocene, must have been exposed, in turn, in what is now Scotland, to deluging outbursts of molten matter from the vexed bowels of the earth, and to overwhelming showers of volcanic ashes.

I would, however, crave attention to the curious fact, that during this immensely protracted period of Plutonic activity, the deep-seated agencies operated in nearly the same lines. Masses of the incarcerated matter seem to have made their escape age after age along the same weak parts of their prison walls,—the earth's crust; and in Scotland we have two of those lines of apparent weakness which converge in a greatly overflown district in the north of Ireland. One of these lines runs along the inner Hebrides nearly south and north, and includes in its area, as distinct centres of Plutonic action, the islands of Skye and of Mull, with what are known as the Small Isles lying between, and the promontory of Ardnamurchan. The other line sweeps across the country from north-east to south-west, commencing at Dunbar on the east, and terminating, in Scotland, with Arran and Campbelton on the west; but running, as I have said, across the Irish Sea, it re-appears in Ulster. It includes, among many lesser trap eminences, the Campsie, the Ochil, and the Lomond hills; the eminences also on which the castles of Stirling and Dumbarton are built; the hills which give character to the scenery around Edinburgh,—Corstorphine, Blackford, the Pentlands, the Castle rock, the Calton, Salisbury Crags, and Arthur Seat; and, far to the east, that Haddington group of trap hills to which North Berwick Law, the Bass, and the Isle of May belong. Beyond these great lines of injected cracks and filled-up craters, especially to the north and east, there are wide districts in Scotland in which there does not occur a single trap rock. The lava-like flood found its way to the surface from the fiery depths beneath, through the chinks and crannies which we now find indicated by the dikes and insulated stacks and hills of what we may term the Lothian and Hebridean lines, and through these only; and those portions of the Lowlands of Scotland which lie to the north of the Grampians, such as the plains of Caithness, Moray, and Easter Ross, present, from the absence of the trap, an entirely different character from that exhibited by the Lowlands of the south.

The igneous rocks have been divided, according to their mineral or mechanical character, into tuffs, amygdaloids, porphyries, dolerites, claystones, clinkstones, wackes, trachytes, and various other species. For our present purpose, however, and as adequate to the demands of our necessarily brief and imperfect sketch, we may regard the trap rocks as consisting of but two great divisions,—first, the traps proper, including all igneous masses, from the porphyries to the basalts, which were ejected from the abyss in a molten form, and which either overflowed from their vents and craters certain portions of the earth's surface, whether subaqueous or subaërial, or, forcing their way between strata of the sedimentary rocks, formed among them dykes, or beds, or pillar-like masses; and secondly, trap-tuffs, which, though igneous in their components, were ejected from craters in the form of loose ashes and detached fragments, or were ground down by the agency of water, and subsequently arranged in regular strata under the same laws which have given their stratification to the rocks of aqueous origin amid which we so frequently find these trap-tuffs intercalated. You will at once see that the division here is a natural one. There is a wide difference betwixt a stratum of broken glass and scoriæ, the débris of a glass-house arranged by the tide on the beach on which it had been cast down a few hours before, and a continuous sheet of plate-glass still retaining its place in the mould into which it had been run off by sluices from the furnace. And such is the difference between trap-tuff and trap proper. We have to arrive, too, when we find them occurring, as in this neighbourhood, among the rocks of a district, at very different conclusions regarding their date and history. Without inquiring whether in some rare instances an eruption of volcanic mud might not possibly be ejected, by a sort of hydraulic-press process, between strata of previously existing rock, and thus a tuff-bed come to be formed which was not only newer than the stratum on which it rested, but also than that by which it was overlaid, we may receive it as a general fact, that the true tuff-bed, like beds of the ordinary sedimentary rocks, is more modern than the stratum on which it rests, and more ancient than the stratum which overlies it; that if it occur, for instance, among the Old Red Sandstones, it belongs to the age of the Old Red Sandstones; if among the Coal Measures, to the age of the Coal Measures; and if among the Oolites, to the age of the Oolites. But we cannot predicate after the same fashion, that the bed of trap proper which we find resting over one series of sedimentary strata and under another is of nearly the same age as the rock above and below, or just a little older than the upper and a little newer than the nether ones. It may have been injected among them many ages after their deposition, during even an entirely different period of the earth's history. We may safely infer, that those beds of stratified trap-tuff which alternate in the Calton Hill with beds of trap-porphyry belong to the Carboniferous period, and are very considerably older than the overlying sandstones and shales on which Regent Terrace is built; but we can no more infer that the great bed of greenstone which forms the picturesque crown of Salisbury Crags is of the same age as the rocks among which it occurs, or, more strictly, a little newer than the strata below and a little older than the strata above, than we can infer that a cast-iron wheel or axle is of the same age as the mould into which it was run, or, more strictly, a little newer than the bottom of the mould, and a little older than the top of it.[1]

Let us now devote a brief space to the consideration of

[1] The usual test of the age of these melted traps is the relation they bear to the rocks which overlie them. If the part of the superjacent bed resting on the igneous rock present an altered appearance, as if it had been more or less baked in a furnace, the trap is regarded as *intrusive*, that is, it forced its way between the planes of the strata, and must consequently be of later age. If, on the other hand, the beds above display no symptom of alteration, and more especially if they consist of trap-tuff, the underlying igneous rock may be presumed to have been erupted either under water or in open air, as the case may be; and hence it is regarded as in a general way contemporaneous with the strata among which it occurs.—G.

the scenery usually associated with the trap rocks,—a subject which should possess some little interest to an Edinburgh audience, seeing that their most magnificent of cities owes almost all that is imposing and peculiar in its aspect and appearance to this cause. The scenery of a trap district may be resolved into two components. In an ancient ruin we frequently see stones hollowed by decay into a sort of fantastic fretwork, not very unlike that which roughens some of our more ancient runic obelisks; and we recognise as the cause of these irregularities of surface on which the effect depends, certain original inequalities in the texture of the mass, and certain weathering influences, which, while they wore away the softer portions, spared such as were harder and more durable. And such, on a larger scale, are the two elements operative in the production of the peculiarities of trap scenery. The hard trap rocks injected into the comparatively soft sandstones and shales of a district, such as that which surrounds the Scottish capital, compose a mass of very various texture and solidity, which, if operated upon *equally* by some power analogous to the weathering one in the case of the fretted stone, would necessarily yield *unequally;* and the weathering influences we find represented on the large scale by the denuding agencies. The noble eminences which give character and individuality to our city were literally scooped out of the general mass by tides, and waves, and deep-acting currents, as the sculptor chisels out his figures, in executing some piece in *alto relievo*, by chipping away the surrounding plane. The bold figure of the poet Hogg becomes almost a literality here :—

> 'Who was it scooped these stony waves?
> Who scalped the brows of old Cairngorm,
> And dug these ever-yawning caves?
> 'Twas I, the Spirit of the Storm.'

The masses of enclosed trap are of various forms. Sometimes they occur as deeply-based pillar-like masses, filling up, it is possible, ancient craters. The rock of hard clinkstone on which the Castle of Edinburgh stands is one of these; but the long inclined plane of sedimentary deposits which it shielded from the wear of the western current interferes with its column-like outline. The Bass Rock is an example of the same kind, with no sedimentary tail to mar the effect of its natural outline. The dike is another and yet more characteristic form of trap rock: it is a rock that was moulded in a longitudinal crack or rent, as the other was moulded in a well-like crater; and when the original matrix in which it was cast has been washed from its sides, and it remains standing up over the level, it assumes the wall-like or *dike-*like form to which it owes its name. In sailing along the west coast of Scotland in a clear sunny day, that gives to each projecting crag its deep patch of shadow, these fragments of walls, of vastly more ancient date than the oldest and most venerable of our Scottish ruins, may be seen rising from the beach along the faces of grassy banks or rounded tuff-formed precipices, and communicating to the general scenery one of its most characteristic features. But one of the main scenic peculiarities of the trap districts is derivable from their trap beds. We find in this neighbourhood, among the hills of the Queen's Park, bed rolled over bed, with bands of shale, or sandstone, or soft trap-tuff, between; and these beds, ranged often in nearly parallel lines, and bared by the denuding agencies, present not unfrequently, seen in profile, the appearance of a flight of steps. Hence the generic name for this class of rocks,—*trappa*, a stair: the traps are the stair-like rocks. As seen in a calm, clear morning, from nearly the eastern termination of Regent Terrace, the Arthur Seat group of hills exhibits three of these beds ranged for considerable distances in nearly parallel lines, and, with these, well-marked fragments of several others. First, reckoning from the west or south, there is the continuous greenstone bed of Salisbury Crags; next, the partially-broken bed of greenstone porphyry known as the Bay Crag; next, the continuous bed of compact greenstone known as the Hill Crag,—that along the top of which the path ascends to the summit of Arthur Seat from St. Anthony's Well; and then there are at least two beds of basalt, partially sanded over, which rise in interrupted steps along the face of the eastern hill. These beds form the peculiar feature of the fine fragment of landscape which from this point of view the Arthur Seat group of hills composes.[1] The trap scenery may be described generally as eminently picturesque. From the circumstance that its eruptive masses rise often from amid level fields, and that its hard abrupt beds, dikes, and columns, alternate often with rich, soft strata, that decompose into fertile soils, it abounds in striking contrasts. The soft plain ascends often at one stride into a hill fantastically rugged and abrupt; and bare and fractured precipices overtop terraced slopes or level platforms, rich in verdure. Some of the more famed scenery of England owes its peculiar beauty to the trap rocks. Hagley, the seat of the Lytteltons, so celebrated in the English poetry of the last century for its beauty, is situated half on a range of picturesque trap hills, half on a level plain of the New Red Sandstone; and the far-famed view from the Leasowes owes much of its beauty to the traps of the Clent Hills. But it would be unpardonable, in treating, however slightly, of the scenery of the trap, to omit all reference to one of its strangest features,—those ranges of polygonal columns which, in at least the more perfect specimens, are peculiar to it, and which impart to Dame Nature, in so many instances, those qualities of proportion and regularity in which art can alone

[1] On the west coast of Mull, and the islands of Gometra and Ulva six or eight of these step-like beds may be seen, rising the one over the other, like terraces or storeys in a building; and the whole landscape seems barred with right lines, that in this district lie nearly parallel to the horizon.

pretend to vie with or surpass her. The specimens in our own neighbourhood are either of small extent, as in Samson's Ribs, or both that and of imperfect form, as at St. Anthony's Chapel and in the adjacent hill-front; but I have seen in the neighbourhood of Linlithgow a range of slender columns sufficiently regular to have given rise to a traditional myth in the locality, that they owe their origin to the ingenuity of the old Picts; and the columned scuir of Eigg greatly surpasses in grandeur the far-famed Giants' Causeway, and scarce falls short of it in the symmetry of its strange architecture. To that wondrous ocean cave of the west which an enlightened age continues to recognise as one of the marvels of Scotland, I need but refer in the graphic verse which the Ettrick Shepherd has transferred, in his *Queen's Wake*, to 'Allan Bawn, the Bard of Mull:'—

> 'Awed to deep silence, they tread the strand,
> Where furnaced pillars in order stand;
> All framed of the liquid burning levin,
> And bent like the bow that spans the heaven;
> Or upright ranged, in wondrous array,
> With purple of green o'er the darkness grey.
> The solemn rows in that ocean den
> Were dimly seen like the forms of men;
> Like giant monks in ages agone,
> Whom the god of the ocean had sear'd to stone;
> And their path was on wondrous pavement old,
> In blocks all cast in some giant mould.'

The *old* scenery of the trap rocks of Scotland,—the scenery associated with them when our country, along at least its two great lines of trappean eruption, was a *Tierra del Fuego*,—a land of fire,—it would require some of that poetic faculty to restore which I would fain challenge for the geologist. Even in the immediate neighbourhood of the capital, the rocky crust of the earth has been heaved into vast waves by the imprisoned Plutonic agencies struggling for vent; huge floods of molten matter, now hardened into mountain masses, have been injected with earthquake throes between the folds of the stony strata; and a submarine volcano has darkened the heavens with its ashes, shutting out during the day the light of the sun, and throwing its red gleam, when the night had fallen, over the steaming eddies of a boiling and broken sea. The area which we now occupy has heaved like the deck of a storm-beset vessel; the solid earth has been rent asunder; and through the wide cracks and fissures, now existing as greenstone dikes, the red molten matter has come rushing through. Could we this evening ascend into the remote past, when that picturesque eminence which overlooks Edinburgh,—according to the poet Malcolm,

> 'Arthur's craggy bulk,
> That dweller of the air, abrupt and lone,'—

was, like the son of Semele, first ushered into the world amid smoke and flame, you would find the scene such as poets might well desire to contemplate, or solicit the aid of their muse adequately to describe. For many ages, what now exists as the picturesque tract of hill and valley attached to old Holyrood, and to which the privileges of the court still extend, had existed as a tract of shallow sea, darkened, when the tide fell, by algæ-covered rocks and banks, and much beaten by waves. From time immemorial has the portion of the earth's crust which underlies that shallow sea been a scene of deep-seated igneous action. Vast beds of trappean rock,—greenstone, and columnar basalt, and amygdaloidal porphyry,—have been wedged from beneath, as molten injections, between the old sedimentary strata; vast waves of translation have come rolling outwards from that disturbed centre, as some submarine hill, elevated by the force of the fiery injection—as the platform of a hydraulic press is elevated when the pump is plied—has raised its broad back over the tide, only, however, to yield piecemeal to the denuding currents and the storm-raised surf of centuries. And now, for day after day has there been a succession of earthquake shocks, that, as the Plutonic paroxysm increases in intensity, become stronger and more frequent, and the mountain-waves roll outwards in ever-widening circles, to rise and fall in distant and solitary seas, or to break in long lines of foam on nameless islands unknown to the geographer. And over the roar of waves or the rush of tides we may hear the growlings of a subterranean thunder, that now dies away in low deep mutterings, and now, ere some fresh earthquake-shock tempests the sea, bellows wildly from the abyss. The billows fall back in boiling eddies; the solid strata are upheaved into a flat dome, crusted with corals and shells; it cracks, it severs, a dark gulf yawns suddenly in the midst; a dense strongly variegated cloud of mingled smoke and steam arises black as midnight in its central volumes, but chequered, where the boiling waves hiss at its edge, with wreaths of white; and anon, with the noise of many waters, a broad sheet of flame rushes upwards a thousand fathoms into the sky. Vast masses of molten rock, that glow red amid even the light of day, are hurled into the air, and then, with hollow sound, fall back into the chasm, or, descending hissing amid the vexed waters, fling high the hot spray, and send the cross circlets of wave which they raise athwart the heavings of the huger billows propelled from the disturbed centre within. The crater rises as the thick showers of ashes descend; and amid the rending of rocks, the roaring of flames, the dashings of waves, the hissings of submerged lava, and the hollow grumblings of the abyss, the darkness of a starless night descends upon the deep. Anon, and we are startled by the shock of yet another and more terrible earthquake; yet another column of flame rushes into the sky, casting a lurid illumination on the thick rolling reek and the pitchy heavings of the wave: seen but for a moment, we may mark the silvery glitter of scales, for there is a shoal of dead fish floating past; and as the coruscations of an electric lightning darts in a thousand fiery tongues from the cloud, some startled monster of the deep bellows in terror from the dank sea beyond.

Let us raise the curtain once more from over the past of the trap districts of Scotland. Myriads of ages have come and gone; the submarine volcano has been long extinguished; and the land, elevated high over the waters, has become a scene of human habitation. But the wild country, marked by the well-known features of abrupt precipitous hill and deep retiring valley, is roughened by many a shaggy wood, and gleams with many a blue lochan, and even its richer plains are but partially broken up by the plough. And lo! the trappean centres of the district are scenes of fierce war, as of old; but it is not the dead uninformed elements,—fire, earth, and water,—but energetic, impassioned man, that now contends, and in fierce warfare battles, with his kind. Yonder, on its trap rock, once the crater of a volcano, is the fortress of the Bass, the stronghold that last surrendered in Britain to William of Nassau; and yonder, on its trap rock, the castle of Dunbar, that brave black Agnes held out in so determined a spirit against the English; and yonder, on its trap rock, the castle of Dirleton, which stood siege in behalf of our country against Edward I.; and yonder, on its trap rock, scaled by Lord Randolph of old when he warred for the Bruce, is the castle of Edinburgh, the scene of a hundred fights, and surrrounded by the halo of a thousand historic associations; and yonder, on its trap rock, is the castle of Stirling, with the battleground of Scotland at its feet, and to maintain which against the greatest of our Scottish kings, the second Edward vainly fought the battle of Bannockburn; and yonder, on its trap rock, is the castle of Dumbarton, long impregnable, but which the soldier of the Reformation won at such fearful risk from the partisans of Mary. I remember at one time deeming it not a little curious that the early geological history of a country should often, as in this instance, seem typical of its subsequent civil history. If a country's geologic history had been much disturbed,—if the trap rock had broken out from below, and tilted up its strata in a thousand abrupt angles, steep precipices, and yawning chasms, I found the chances as ten to one that there succeeded, when man came upon the scene, a history, scarce less disturbed, of fierce wars, protracted sieges, and desperate battles. The stormy morning during which merely the angry elements had contended, I found succeeded in almost every instance by a stormy day maddened by the turmoil of human passion. But a little reflection dissipated the mystery; though it served to show through what immense periods mere physical causes may continue to operate with moral effect, and how, in the purposes of Him who saw the end from the beginning, a scene of fiery confusion,—of roaring waves and heaving earthquakes, of ascending hills and deepening valleys,—may have been closely associated with the right development and ultimate dignity and happiness of the moral agent of creation,—unborn at the time,—reasoning, responsible man. It is amid these centres of geologic disturbance, the natural strongholds of the earth, that the true battles of the race, the battles of civilisation and civil liberty, have been successfully maintained by handfuls of hardy men, against the despot-led myriads of the plains. In glancing over a map of Europe and the countries adjacent, on which the mountain groups are marked, you will at once perceive that Greece and the Holy Land, Scotland and the Swiss cantons, formed centres of great plutonic disturbance of this character. They had each their geologic tremors and perturbations,—their protracted periods of eruption and earthquake,—long ere their analogous civil history, with its ages of convulsion and revolution, in which man was the agent, had yet commenced its course. And, indirectly at least, the disturbed civil history was in each instance a consequence of the disturbed geologic one.

From the Tertiary deposits we pass direct to the few scattered remains which survive in Scotland of the Cretaceous period. It is now nearly thirty years since it was found by geologists that chalk flints, enclosing in many specimens the peculiar organisms of the system, occur in the superficial deposits of Banff and Aberdeenshires; and about three years ago they were also discovered by a very ingenious man, a Thurso tradesman, Mr. Robert Dick, in the boulder-clays of Caithness. It is, however, a curious fact, that what the geologist has only come to know within the course of the present generation was well known to the wild aboriginal inhabitants of the country some three or four thousand years ago. Well-nigh one half the ancient arrow and smaller javelin heads of the stone period in Scotland, especially those found to the north of the Grampians, were fashioned out of the yellow Aberdeenshire flints. A history of those arts of savage life which the course of discovery served to supplant and obliterate, but which could not be carried on without a knowledge of substances and qualities afterwards lost, until re-discovered by scientific curiosity, would form an exceedingly curious one. On finding, a good many years ago, a vein of a bituminous jet in one of the ichthyolite beds of the Old Red Sandstone of Ross,— beds unknown at the time to even our first geologists,—it curiously impressed me to remember that my discovery was, after all, only a discovery at second-hand; for that in an unglazed hand-made urn of apparently a very early period, dug up in the neighbourhood only a few years before, there had been found a very primitive necklace, fashioned out of evidently the same jet. It would seem that to these ichthyolite beds, unknown at the time in the district to all but myself, the savage inhabitants had had recourse for the materials of their rude ornaments thousands of years before.

They were mineralogists enough, too, as their stone hatchets and battle-axes testify, to know where the best tool-and-weapon-making rocks occur; and I once found in a northern locality a battle-axe of an exceeding strong and tough variety of indurated talc, that nearly approached in character to the axe-stone of Werner, which, if native to Scotland at all, is so in some primary district which I am not mineralogist enough to indicate. It shows us after how strange a fashion extremes may meet,—that rude savages, ignorant of the use of the metals, and the scientific explorers of a highly civilized age, rationally desirous to know how the adorable Creator wrought upon this earth of old, ere man had yet entered upon it as a scene of probation, should have formed an acquaintance with the same classes of objects,—classes of objects of which the men of an intervening period knew nothing.

The chalk fragments and flints of Caithness and Banff seem to have been carried eastwards on the occidental current of the Pleistocene period,—those of the one county from that western portion of the chalk ring or girdle to which I have already referred as lying in the Atlantic, and those of the other from that eastern portion of the ring which is buried in the outer reaches of the Moray Firth. In Aberdeenshire, however, some twenty miles or so to the north of the city, in the parish of Ellon and some of the contiguous parishes, and running at a considerable distance inland in a line nearly parallel to the coast, the flints so abound, and, unlike those of the English gravels, are so little water-worn, as to give evidence that they must have been derived from the disintegration of outliers of the system that once existed, it is probable, in their immediate neighbourhood. They overlie, too, in some parts of this locality, what seems to be a re-formation of the greensand; of which the soft incoherent masses, containing, as they do, in some instances in a good state of keeping, some of the more

H

fragile organisms of the deposit, could not possibly have travelled far. The fossils of our chalk flints and of the underlying greensand are sufficiently numerous and characteristic to serve the purpose of identifying the worn and scattered deposits in which they occur with the amply developed chalks and greensands of England, but perhaps not sufficiently so, nor yet always in a sufficiently fine state of preservation, to render the district a very hopeful scene of labour to the collector desirous absolutely to extend our knowledge of the extinct forms of life. I have seen, however, especially in the collections of Dr. Fleming, the Rev. Mr. Longmuir of Aberdeen, and Mr. Fergusson of Glasgow, fine and very characteristic specimens of the Scotch Chalk,—delicate flustra sponges and corals locked up in flint,—well-marked portions of the sea-egg order (Echinidæ) belonging to the cidarite, galerite, and spatangus families,—terebratulæ of various species,—good specimens of that very characteristic conchifer of the Chalk, the Inoceramus,—with casts of minute belemnites and portions of ammonites and baculites. The group of remains preserved is unequivocally that of the Cretaceous fauna, just as Scotland has also a group of archæological remains decidedly Roman; though in either case these remains serve but for purposes of identification with larger groups elsewhere; and in order thoroughly to study either the one or the other, the antiquary or geologist would have to remove from what is equally the outskirts of the old Roman or old Cretaceous empire, towards its centre in the south.

All our geologists agree in holding that the Chalk was deposited in an ocean of very considerable depth, and of such extent that it must have covered for many ages the greater part of what is now southern and central Europe. It has been traced in one direction from the north of Ireland to the Crimea in Southern Russia, a distance of about twelve hundred miles, and in another direction from the south of Sweden to the south-west of France, a distance of about nine hundred miles; and there are extensive districts both in France and England where it attains to an average thickness of not less than a thousand feet. The only analogous deposit of the present time occurs on comparatively a small scale among the coralline reefs and lagoons of the Pacific, where there is in the act of forming an impalpable white mud derived from the corals, which in dried specimens cannot be distinguished by the unassisted eye from masses of soft chalk. But what chiefly distinguishes the true chalk from any of its modern representatives is the amazing number of microscopic animals which it contains. On a low estimate half its entire bulk is composed of animalculites of such amazing minuteness, that it has been calculated by Ehrenberg that each cubic inch of chalk may contain upwards of a million of the shells of these creatures. The chalk rocks so characteristic of the sister kingdom have been often sung by the poets as

'Rising like white ramparts all along
The blue sea's border.'

And, in especial, one 'chalky bourn of dread and dizzy summit' has been made by the greatest of poets the subject of the sublimest description of a giddy, awe-inspiring precipice ever drawn. And here is there a new association with which to connect the chalk cliffs of England. Every fragment of these cliffs was once associated with animal life; that impalpable white dust which gives a milky hue to the waves as they dash against them, consists of curiously organized skeletons; even the white line which I draw along the board, were our eyes to be suddenly endowed with a high microscopic power, would resemble part of the wall of a grotto covered over with shells. And, embedded in this mass of minute, nicely-framed invisibilities,—Polythalamia, Foraminifera, Polyporia, and Diatomaceæ,—we find fossils of larger size, such as *Spatangus-cor* and the spiny *Plagiostoma*, which seem to have found proper habitats in the mud formed by the dead remains of these animalculæ. Curious examples of a similar kind may be still seen among the Hebrides, of sand-burrowing molluscs and echinoderms finding habitats amid accumulations of the débris of organic life, chiefly comminuted shells, on coasts where otherwise there could have been no place for them. The deep-sea shells propelled shorewards by the agency of tides and waves are ground down by the action of the surf against the rocks. They may be seen occurring in the hollows of the skerries, as one passes shorewards along some of the rocky bays, in handfuls of more and more comminuted fragments, just as, in passing along the successive vats of a paper-mill, one finds the linen rags more and more disintegrated by the cylinders; and then, within some sheltering shelf or ledge, we find the gathered handfuls of former ages spreading into a wave-rippled beach of minute shelly particles, that presents, save in its snow-white colour, the appearance of sandy beaches of the ordinary mineral components. But the beach once formed in this way soon begins to receive accessions from the exuviæ of animals that love such localities,—spatangi, razor-fish, cockles, and the several varieties of the gaper family,—and that enjoy life agreeably to their natures and constitutions, not in the least saddened by the idea that they are living amid the rubbish of a charnel-house; and sometimes one-half the whole beach comes thus to be composed of a class of remains that, save for the previous existence of the other half of it, could not have been formed in such localities at all. Now, such must have been the state of matters in the times of the Chalk. Unnumbered millions must have died in order that the medium might be provided in which a class of their successors could alone live. Of the land which skirted this ocean of the Chalk, or of its productions, we know almost nothing. There have been found in Chalk flints a few fragments of silicified wood, and, in one or two instances, the cones of cycadaceous plants; and the upper beds of the system have furnished the remains of a gigantic lizard,—the Mosasaurus, with those of turtles, tortoises, and Pterodactyls. True, the Mosasaurus *may* have been, as Cuvier supposed, a marine reptile, and the turtles *must* have been so; but then both, as egg-bearing animals, must have brought forth their young on some shore; and the tortoises, with the Pterodactyl or flying lizard, must be regarded as decidedly terrestrial. Such is almost all we yet know of the flora or fauna of the land of the Chalk; whereas in marine organisms the system is so exceedingly rich, that its ascertained species amount, we find it stated by Brown, to about three thousand. The geologic diorama abounds in strange contrasts. When the curtain last rose upon our country, we looked abroad over the amber-producing forests of the Tertiary period, with their sunlit glades and brown and bosky recesses, and we saw, far distant on the skirts of the densely wooded land, a fire-belching volcano, over-canopied by its cloud of smoke and ashes. And now, when the curtain again rises, we see the same tract occupied, far as the eye can reach, by a broad ocean, traversed by a pale milky line, that wends its dimpling way through the blue expanse, like a river through a meadow. That milky way of turbid water indicates the course of a deep-setting current, that disturbs, far beneath, the impalpable mud of the Chalk. Sailing molluscs career in their galleys of pearl over the surface of this ancient sea; fishes of long extinct species dart with sudden gleam through its middle depths; and far below, on its white floor, the sea-urchin creeps, and the spatangus burrows, and crania and terebratulæ have cast anchor, and the *Crista Galli* (or carinated oyster) opens its curiously plicated valves, carved with the zigzag mouldings of a Norman doorway, and the flower-like marsupite expands its living petals. And, dim and distant in the direction of the future Grampians, we may espy a cloud-enveloped island; but such is its remoteness, and such the enveloping haze, that we can know little more than that it bears along its shores and on its middle heights a forest of nameless trees, unchronicled by the fossil botanist.

In bringing to a close this part of my subject, let me here remark that, if we except the obscure and humbly organized diatomaceæ,—a microscopic family of organisms which some of our authorities deem animal and some vegetable, and of which hundreds and thousands would find ample room in a single drop of water,—we have now reached a point in the history of our country in which there existed no *species* of plant or animal that exists at the present time. Not a reptile, fish, mollusc, or zoophyte of the Cretaceous system continues to live. We know that it is appointed for all individuals once to die, whatever their tribe or family, because hitherto all individuals *have* died; and Geology, by extending our experience, shows us that the same fate awaits on species as on the individuals that compose them. In the one case, too, as in the other, death has its special laws; but the laws which determine the life and death of species seem widely different from those which regulate the life and death of individuals and generations. In general, and with but a few exceptions in favour of the cold-blooded division of the vertebrata, the higher orders of animals live longest. A man may survive for a hundred years; an ephemera bursts from its shell in the morning, and dies at night. But it is far otherwise with the higher orders of species. Molluscs and corals outlive the vertebrata; and tribes of the low infusory animals outlive molluscs and corals. We know not that a single shell of at least the latter Pleistocene period has become extinct; but many of its noblest quadrupeds, such as the Irish elk, the cave-bear, tiger, and hyena, and the northern rhinoceros, hippopotamus, and elephant, exist no longer. And as we rise into the remote past, and take farewell, one after one, of even the lower forms,—shells and corals,—and get into a formation all of whose visible organisms are old-fashioned and extinct, we apply the microscope to its impalpable dust, and again, among still humbler and lowlier shapes, find ourselves in the presence of the familiar and the recent. In another sense than that which the old poet contemplated, we learn from the history of species that the most lowly are the most safe.

> ' The tallest pines feel most the power
> Of wintry blasts ; the loftiest tower
> Comes heaviest to the ground.
> The bolts that spare the mountain side
> His cloud-capt eminence divide,
> And spread the ruin round.'

How long some of these extinct species may have lived we know not, and may never know; but in all cases their term of existence must have been very extended. Even the extinct elephant lived long enough as a species to whiten the plains of Siberia with huge bones, and to form quarries of ivory that have furnished the ivory market for year after year with its largest supplies. And of some of the humbler species of animals the period during which they have continued to live must have been vastly more protracted. *Cyprina Islandica* seems to have come into existence at least as early as the fossil elephant; and now, thousands of years after the boreal pachyderm is gone, the boreal shell still exists by millions, and evinces no symptom of decline. And yet, since the commencement of the great Tertiary division, series of shells, as hardy, apparently, as Cyprina, have in succession come into being, and then ceased to be. The period over which we have passed includes *generations* of species. But there was space enough for them all in the bygone eternity. It has sometimes appeared to me as if, from our own weak inability to conceive of the upper reaches of that awful tide of continuity which had no beginning, and of which the measured shreds and fragments constitute time, we had become jealous lest even God himself should have wrought in it during other than a brief and limited space, with which our small faculties could easily grapple.

> 'Oh, who can strive
> To comprehend the vast, the awful truth
> Of the eternity that hath gone by,
> And not recoil from the dismaying sense
> Of human impotence! The life of man
> Is summed in birthdays and in sepulchres,
> But the eternal God had no beginning.'

There are two great infinites,—the infinite in space and the infinite in time. It were well, surely, to be humble enough to acknowledge it accordant to all analogy, that as He who inhabits eternity has filled the one limitless void—that of space—with world upon world and system upon system, far beyond the reach of human ken, He should also have wrought in the other limitless world—that of time—for age after age, and period after period, far beyond the reach of human conception.

LECTURE FOURTH.

The Continuity of Existences twice broken in Geological History—The three great Geological Divisions representative of three independent Orders of Existences—Origin of the Wealden in England—Its great Depth and high Antiquity—The question whether the Weald Formation belongs to the Cretaceous or the Oolitic System determined in favour of the latter by its Position in Scotland—Its Organisms, consisting of both Salt and Fresh Water Animals, indicative of its Fluviatile Origin, but in proximity to the Ocean—The Outliers of the Weald in Morayshire—Their Organisms—The *Sabbath-Stone* of the Northumberland Coal Pits—Origin of its Name—The Framework of Scotland—The Conditions under which it may have been formed—The Lias and the Oolite produced by the last great Upheaval of its Northern Mountains—The Line of Elevation of the Lowland Counties—Localities of the Oolitic Deposits of Scotland—Its Flora and Fauna—History of one of its Pine Trees—Its Animal Organisms—A Walk into the Wilds of the Oolite Hills of Sutherland.

THE mystic thread, with its three strands of black, white, and grey, spun by the sybil in *Guy Mannering*, formed, she said, a 'full hank, but not a haill ane :' the lengthened tale of years which it symbolized ' was thrice broken and thrice to asp.' I have sometimes thought of that wonderfully mingled and variously coloured thread of existence which descends from the earliest periods known to the geologist down to our own times, as not unaptly represented by that produced on this occasion from the spindle of the gipsy. We find, in its general tissue, species interlaced with and laying hold of species, as, in the thread, fibre is interlaced with and lays hold of fibre ; and as by this arrangement the fibres, though not themselves continuous, but of very limited length, form a continuous cord, so species of limited duration, that at certain parts in the course of time began to be, and at certain other parts became extinct, form throughout immensely extended periods a continuous cord of existence. New species had come into being ere the old ones dropped away and disappeared; and there occurred for long ages no break or hiatus in the course, just as in the human family there occurs no abrupt break or hiatus, from the circumstance that new generations come upon the stage ere the old ones make their final exit. But in the geological thread, as in that of the sybil, the continuity is *twice* abruptly broken, and the thread itself divided, in consequence, into three parts. It is continuous from the present time up to the commencement of the Tertiary period; and then so abrupt a break occurs, that, with the exception of the microscopic diatomaceæ, to which I last evening referred, and of one shell and one coral, not a single species crosses the gap. On its further or remoter side, however, where the Secondary division closes, the intermingling of species again begins, and runs on till the commencement of this great Secondary division; and then, just where the Palæozoic division closes, we find another abrupt break, crossed, if crossed at all,—for there still exists some doubt on the subject,—by but two species of plant.[1] And then, from the further side of this second gap the thread of being continues unbroken, until we find it terminating with the first beginnings of life upon our planet. Why these strange gaps should occur,—why the long descending cord of organic existence should be thus mysteriously broken in three,—we know not yet, and never may; but, like the division into books and chapters of some great work on natural history, such as that of Cuvier or Buffon, it serves to break up the whole according to an intelligible plan, the scheme of which we may, in part at least, aspire to comprehend. The three great divisions of the geologist,—Tertiary, Secondary, and Palæozoic,—of which these two chasms, with the be-

[1] For a reference to the research of the last two years, which has been busily at work upon this precise epoch, see Preface.

ginnings of life on the one hand, and the present state of things on the other, form the terminal limits,—represent each, if I may so express myself, an independent dynasty or empire. Under certain qualifications, to which I shall afterwards refer, the Tertiary division represents the dynasty of the mammal; the Secondary division the dynasty of the reptile; and the Palæozoic division the dynasty of the fish. Each of the divisions, too, has a special type or characteristic fashion of its own; so that the aspect of its existences differs as much in the group from the aspect of the existences of each of the others, as if they had been groups belonging to different planets. The vegetable and animal organisms of the planet Venus may not differ more from those of the planet Mars, or those of Mars from the organisms of the planet Jupiter, than the existences of the Tertiary division differ from those of the Secondary one, or those of the Secondary one from the existences of the Palæozoic division.

Beneath the two great divisions of the Cretaceous system, and consequently of more ancient date, there occurs in the sister kingdom an important series of beds, chiefly of lacustrine or fluviatile origin, known as the Wealden. Before the submergence of what are now the south-eastern parts of England, first beneath the comparatively shallow sea of the Greensand, and then beneath the profounder depths of the ocean of the Chalk, a mighty river, the drainage of some unknown continent, seems to have flowed for many ages along those parts of Kent, Surrey, and Sussex, known as the Valley of the Weald. The banks of this old nameless river were covered with forests of coniferous trees of the Pine and Araucarian families, with cycadeæ and ferns, and were haunted by gigantic reptiles, herbivorous and carnivorous, some of which rivalled in bulk the mammoth and the elephant; its waters were inhabited by amphibiæ of the same great class, chiefly crocodiles and chelonians of extinct species and type; by numerous fishes, too, of the old ganoid order; and by shells whose families, and even genera, still exist in our pools and rivers, though the species be all gone. Winged reptiles, too, occasionally flitted amid its woods, or sped over its broad bosom; and insects of the same family as that to which our dragon-flies belong spent the first two stages of their existence at the bottom of its pools and shallows, and the terminal one in darting over it on their wings of delicate gauze in quest of their prey. It is stated by Dr. Mantell, our highest authority on the subject of the Weald, that the delta of this great river is about two thousand feet in thickness,—a thickness which quadruples that of the delta of the Mississippi. There can be little doubt that the American 'Father of Waters' is a very ancient river; and yet it would seem that this river of the Wealden, which has now existed for myriads of ages in but its fossilized remains, hidden under the Wolds of Surrey and Kent,—this old river, which flowed over where the ocean of the Oolite once had been, and in turn gave place and was overflowed by the ocean of the Chalk,—continued to roll its downward waters amid forests as dense and as thickly inhabited as those of the great American valley, during a period perhaps four times as extended.

Compared with the English formation of the Weald, which extends over a wide, and what was at one time a very rude district, our beds of the Scotch Wealden are but of little depth, and limited extent. And yet they serve to throw a not unimportant light on the true character and place of the formation. It occurs in England, as I have said, between two great marine systems,—the Cretaceous and the Oolitic; and the question has arisen, to which of these systems does it belong? Now, our Scotch beds of the Weald determine the question. They make their appearance, not at the top of the Oolitic deposits, as in England, but intercalated throughout the system,—occurring in the Isle of Skye, where they were first detected many years ago by Sir Roderick Murchison, immediately under the Oxford clay, a bed of the Middle Oolite; and at Brora, where they were first detected a few twelvemonths since by Mr. Robertson of Elgin, in pretty nearly the same medial position, and where what is known as the great Oolite occurs. Three years ago I had the pleasure of detecting a bed of the same lacustrine or estuary character, and bearing many of the characteristic marks of the Weald, greatly lower still,—lower, indeed, than any fresh-water deposit of the Secondary division in Britain. I found it occurring not forty yards over the bottom of the Lias,—the formation which constitutes the base of the Oolitic system. In Morayshire the Weald occurs in the form of outliers, that rise, as at Linksfield, in the immediate neighbourhood of Elgin, into low swelling hills, resting on the Old Red Sandstone of the district, and so thoroughly insulated from every other rock of the same age, that they have reminded me of detached hillocks of débris and ashes shot down on the surface of some ancient moor by some painstaking farmer, who had contemplated bringing the waste under subjection to the plough. But though valueless, from their detached character, for determining the place of the formation, they serve better than the intercalated beds of Ross, Skye, and Sutherland, to establish by their animal remains the palæontological identity of the Scotch with the English Wealden.

Rather more than twelve years ago, the late Dr. John Malcolmson of Madras,—a zealous and accomplished geologist, too early lost to science and his friends,—brought with him, when on a visit to the Continent, several specimens of ichthyic remains from a Morayshire deposit, and submitted them to Agassiz. 'Permit me,' said the naturalist, 'to find out for myself the formation to which they belong.' He passed hand and eye over tooth and spine, plate and bone, and at length set his finger on a single scale of rhomboidal form and brightly enamelled surface. 'Some of these teeth,' he said, 'belong to the genus Hybodus, but the species are new, and the genus itself has a wide range. Here, however, is something more determinate. This scale belongs to the *Lepidotus minor*, or ichthyolite of the Weald, and one of the most characteristic fishes of the great fresh-water formation of Surrey and Kent.' The fossils on which the distinguished ichthyologist thus promptly, and, as it proved, correctly decided, had been collected by Dr. Malcolmson from the the Wealden outlier at Linksfield ; and the ichthyolite which he so specially singled out,—the Lepidotus,—seems to have been a fresh-water fish of the nearly extinct ganoid order, and more nearly akin to the Lepidosteus of the North American rivers and lakes than to any other fish that now exists. By much the greater number of its contemporaries in the deposit also belonged to lakes and rivers. Some of the limestone slabs are thickly covered over by fresh-water shells, of types very much akin to those which still occur in our pools and ditches, such as Planorbis and Paludina. It presents also beds of a fresh-water mussel akin to a mussel of the English Weald,—*Mytilus Lyellii;* and it so abounds in the remains of those minute, one-eyed crustaceans known as the Cyprides, that the vast numbers of their egg-shaped shelly cases give to some of the beds a structure resembling the roe of a fish. It contains, too, bones of a species of tortoise, and several other decidedly fresh-water remains ; while another class of its organisms serve to show that it was occasionally visited by denizens of the sea. It has furnished specimens of bones and teeth of Plesiosaurus,—a marine reptile ; and some of the upper beds contain a small oyster ; while a class of its remains,—the teeth and huge dorsal spines of Hybodonts, an extinct family of sharks, though they may have been fitted to sustain life in brackish water, seem to indicate rather a sea than a lacustrine or river habitat. The deposit took place in all probability in the upper reaches of an estuary operated upon by the tides, and at one time fresh and at another brackish, and where, in a certain debatable tract, the fishes, reptiles, and shells of the river met and mingled with the fishes, reptiles, and shells of the sea. I may mention, that in the immediate neighbourhood of the fresh-water or Weald beds, intercalated, as in Ross and Sutherland, with the marine deposits of the Lias or Oolite, there always occur beds of a species of shell, which, though it exhibits internally a peculiar structure of hinge, unlike any other known to the conchologist, bears externally very much the appearance of a mytilus or mussel. It seems to have lived in brackish water, and to have marked a transition stage between the marine and lacustrine, —the salt and the fresh; for immediately under or over it, as the case occurs, the explorer is ever sure to find productions of the land or of fresh water,—lake or river shells, such as cyclas or paludina, or portions of terrestrial plants, and occasionally of fresh-water tortoises. This transition shell is known as the Perna. These notices you will, I am afraid, deem tediously minute; but they indulge us with at least a glimpse of a portion of what is now our country during an immensely extended period, of which no other record exists. Where some nameless river enters the sea, we determine, as through a thick fog, which conceals the line of banks on either hand, that the waters swarm with life, reptilian and ichthyic: the glossy scales of the river *Lepidotus* gleam bright through the depths; while the shark-like *Hybodus* from the distant ocean shows above the surface his long dorsal fin, armed with its thorny spine; and over beds of shells of mingled character, a carnivorous fresh-water tortoise, akin to the fierce Trionyx of the southern parts of North America, meets with the scarce more formidable sea-born Plesiosaurus.

In these Morayshire outliers of the Weald we first find *in situ* in our country (for we need scarce take into account the Tertiary beds of Mull), fossiliferous deposits that have been converted into solid rock; and certainly the appearance of some of the sections is such as to awaken curiosity. In the section of Linksfield, in the neighbourhood of Elgin, though the thickness of the deposit does not exceed forty feet, there occur numerous alternations of argillaceous and calcareous beds, differing from each other in colour and quality, and not unfrequently in their fossils also; and each of which evidently represents a state of things which obtained during the period of their deposition, distinct from the preceding and succeeding states.[1] Strata of grey, green, blue, and almost black clays, alternate with beds of light green, light brown, grey, and almost black limestones; and such is the effect, when a first section is opened in the deposit, as sometimes happens to facilitate the working of a limestone quarry below, that one is reminded, by the variety and peculiar tone of the colours, of the inlaid work of an old-fashioned cabinet made of the tinted woods which were in such common use about two centuries ago. Some of these bands seem, from their contents, to be of fresh water; some of marine origin; one bed nearly four feet in thickness is composed almost exclusively of the shelly coverings of a minute crustacean,—*Cypris globosa,*—not half the size of a small pinhead; one is strewed over with the teeth of sharks; one with the plates and scales of ganoidal fishes; in one a small mussel is exceedingly abundant; another contains the shells of Planorbis and Paludina; in this layer we find a small

---

[1] Fielding, in his *Voyage to Lisbon* (1754), gives an account of an inaccessible bank of mud which stretched at low water between the shore at Ryde and the sea. 'Between the shore and the sea,' he says, 'there is at low water an impassable gulf of deep mud, which can neither be traversed by walking nor swimming, so that for near one-half of the twenty-four hours Ryde is inaccessible by friend or foe.' The same tract now is occupied by an expanse of firm white sand, which forms excellent bathing ground; but immediately under, at the depth of from eighteen inches to two feet, the mud of Fielding's days is found occurring as a dark-coloured impalpable silt.

oyster, which must have lived in the sea; in that, a Cyclas, the inhabitant of a lake; here the plates of a river tortoise; there the bones of the marine Plesiosaur. Of all the many-coloured strata of which the deposit consists, there is not one which does not speak of that law of change of which the poet, as if in anticipation of the discoveries of modern science, sings so philosophically and well:—

> 'Of chance or change, oh! let not man complain,
> Else shall he never, never cease to wail;
> For from the imperial dome, to where the swain
> Rears the lone cottage in the silent dale,
> All feel the assault of Fortune's fickle gale;
> Art, empire, earth itself, to change are doom'd:
> Earthquakes have raised to heaven the humble vale,
> And gulfs the mountain's mighty mass entomb'd;
> And where the Atlantic rolls, wide continents have bloom'd.'

Regarded, too, as the record of, if I may so express myself, a party-coloured time, these party-coloured layers are of no little interest. There forms in the recesses of the Northumbrian coal-pits a party-coloured clay, consisting of grey and black layers, which, from a certain peculiarity to which I shall immediately advert, bears the name of *Sabbath-stone*. The springs which ooze into the pits are charged with a fine impalpable pipe-clay, which they deposit in the pools and waters of the deserted workings, and which is of a pale grey colour approaching to white. When the miners are at work, however, a light black dust, struck by their tools from the coal, and carried by currents of air into the recesses of the mine, is deposited along with it; and, in consequence, each day's work is marked by a thin black layer in the mass, while each night, during which there is a cessation of labour, is represented by a pale layer, which exhibits the colour natural to the clay. And when a cross section of the substance thus deposited comes to be made, every week of regular employment is found to be represented by a group of six black streaks closely lined off on a pale ground, and each Sabbath by a broad pale streak interposed between each group,—exactly such a space, in short, as a clerk, in keeping tally, would leave between his fagots of strokes. In this curious record a holiday takes its place among the working days, like a second Sabbath. 'How comes this week to have two Sabbaths?' inquired a gentleman to whom a specimen was shown at one of the pits. 'That blank Friday,' replied the foreman, 'was the day of the races.' 'And what,' said the visitor, 'means this large empty space, a full fortnight in breadth and more!' 'Oh, that space,' rejoined the foreman, 'shows the time of the strike for wages: the men stood out for three weeks, and then gave in.' In fine, the Sabbath-stone of the Northumbrian mines is a sort of geologic register of the work done in them,—a sort of natural tally, in which the sedimentary agent keeps the chalk, and which tells when the miners labour and when they rest, and whether they keep their Sabbaths intact or encroach upon them. One would scarce expect to find of transactions so humble a record in the heart of a stone; but it may serve to show how very curious that narrative might be, could we but read it aright, which lies couched in the party-coloured layers of the Morayshire Wealden. All its many beds, green, black, and grey, argillaceous and calcareous, record the workings of nature, with her alternations of repose, in a time of frequent vicissitude, and amid its annals of chemical and mechanical change embodies in many an episodical little passage its exhibitions of anatomical structure and its anecdotes of animal life.

Before passing on to the Oolite, as developed in Scotland, or rather to our Scotch deposits of the marine Oolite, —for what we call our Wealden is, as I have shown, merely an estuary or lacustrine Oolite,—let me solicit your attention to a few points illustrative of what may be termed the framework of our country. There are two sets of conditions under which land may arise from the ocean. Its hills and plateaus may be formed by the subterranean forces violently thrusting them up, like vast wedges, through the general crust of the earth, and high over the ocean level; or it may be brought up to the light and air *en masse* by a general elevation over wide areas of the unbroken crust itself; or land may again sink under these two sets of conditions: it may sink in consequence of a breaking up and prostration of its framework to the average level of the crust,—of a striking back, if I may so speak, of the protruded wedges; or it may sink in consequence of a *general depression* over a wide area of the portion of the crust on which its framework is erected. Thus Scotland might disappear under the waves, either by some violent earthquake convulsion that would strike down its hills and table-lands to the general level of the earth's crust, and of consequence wholly destroy its contour; or it might disappear through a gentle sinking of the area that it occupies, which would leave its general contour unchanged. Were there a depression to take place where it now rises, of but one foot in five hundred over an area a thousand miles square, its highest mountain-summits would be buried beneath the sea, and yet the contour of the submerged land would remain almost identically what it is,—its hills would retain the same relative elevation over its valleys, and its higher table-lands over its lower plains. Now, in the later ages of its history,—in those ages, for instance, in which the ice-laden ocean of the boulder-clay rose high along its hill-sides, and it existed as a wintry archipelago of islands, there seems to have taken place scarce any change in its framework: the depressions through which it sank, and the elevations through which it rose, seem to have been depressions and elevations of area; and, whether under or over the waves, it continued to retain its general contour. The last great change which affected its framework, and gave to it a different profile in relation to the general surface of the globe from that which it had borne in the earlier ages,—the change which thrust up its latest-born lines of mountains like wedges through the earth's crust,—was a change which took place a little posterior to that period of its history at which I am now arrived. We find that its last lines of hills disturbed and bore up with them deposits of the Lias and of the Oolite, but of no later formation. The gigantic Ben Nevis and his Anakim brethren of the same group were raising their heads and shoulders through the earth's crust, to form the future landmarks of our country, shortly after the period when the river Lepidoids of the Wealden were disporting in the same brackish tract with the Hybodont sharks of its seas, and its fresh-water Chelonians and marine Plesiosauri met and intermingled in the same neutral rocks of estuary.

The last great paroxysm of upheaval among our Scottish mountains seems to have operated in lines that traversed the country diagonally from nearly south-west by south to north-east by north,—the line indicated by that of the great Caledonian Valley. We find a northern district of considerable extent ploughed in this direction by the great parallel glens traversed by the Spey, the Findhorn, the Nairn, and the Ness. The northern shore of the Moray Firth, too, with that remarkable line of hills which includes the Sutors of Cromarty, pertains to this system, as also the higher mountain range which rises along the coast of Sutherland, and to which the Ord Hill of Caithness belongs. These lines of hills, wherever they have come in contact—as along the shores of the Moray Firth—with beds of the Lias and Oolite, have disturbed and tilted up, at a steep angle, their edges. The hill of Eathie, in the neighbourhood of Cromarty,—a hill of the series in which the two Sutors occur,—has at one place borne up the Lower Lias on its flanks at an angle of eighty; and among the rocks of the Northern Sutor there is a tall precipice of the Old Red Sandstone, with an uptilted deposit of the Lias at its base, whose abrupt, dizzy front, once the haunt of the eagle, and still that of the blue hawk, was evidently, ere the elevation of the series, part of the horizontal platform on which the first Liassic stratum had been deposited. What was a flat submarine bottom then is a steep ivy-mantled precipice now. Across the long deep valleys and mountain ridges of this last line of upheaval in Scotland,—the line to which Ben Nevis, Milfourveny, and the Ord Hill of Caithness belong, and whose period of elevation a high Continental authority, Elie de Beaumont, regards as identical with that of the Mont Pilas and Côte d'Or of France, we find a greatly less continuous, because more interrupted and broken, set of ridges, running in a nearly westerly direction. The firths of Dornoch, Cromarty, and Beauly, with the bays of Munlochy and Urquhart, Loch Oich and Loch Eil, which all strike westwards across the country from off the great diagonal trench of the Caledonian Valley, indicate the direction of this second and earlier line of upheaval. I say earlier line. The hills of the diagonal Ben Nevis line disturbed and broke up the Oolite, whereas the hills of the transverse, or, as I may term it, Ben Wyvis line, disturbed and bore up with them nothing more modern than the Old Red Sandstone. I have described the northern part of the kingdom as consisting of a great Primary nucleus, surrounded by strata more or less broken, of Old Red Sandstone, Lias, and Oolite.[1] Let us now further conceive of that nucleus as a stony field, that had been first ploughed across and fretted into deep furrows and steep mountainous ridges, and then in an after period ploughed diagonally, so as partially to efface the former ploughing, so that only in the direction of the last ploughing do the ridges and furrows remain tolerably entire, —let us, I say, conceive of such a ploughed field, and we shall have a tolerably adequate conception, so far as it goes, of the framework of at least the northern portion of Scot-

[1] To which is to be now added Silurian.

land. In the southern part of the kingdom there is yet another line of elevation exhibited, whose direction from nearly north-east to south-west we find indicated by the nearly parallel lines in which the greater formations of the Lowland counties, from the clay-slates that flank the Grampians, to the Grauwackes of the border districts, sweep across the country. I fear that the homely illustrations which I have to employ in rendering my subject comprehensible,—such as wedges struck upwards from below,—a field first ploughed across and then diagonally,—may have the effect of so reducing my subject in your minds into a mere model, that, through the necessary reduction, more may be lost in expansiveness of feeling than gained by any substitution of clearness of view. There can be little doubt that in the conceptions of mind, as in the collocations of matter, the *portable* means the small; and that Goethe exercised his wonted shrewdness in remarking, that when the ancients spoke of the unmeasurable earth and the illimitable sea, it was with a profounder feeling than any now exercised by the geographer in a time when every school-girl can tell that the world is round. You will, however, remember, that though my illustrations are small, my subject is large; and such of my audience as have sailed over the profound depths of Loch Ness,—depths greatly more profound than those of the German Ocean beyond,—and seen those lines of russet mountains, so often capped with cloud, and so often, even at midsummer, streaked with snow,—that rise on either hand, and that enclose from sea to sea that mighty trench which the old unsophisticated Highlander learned to distinguish as the great Glen of Albyn,—when they call up to memory the noble features of the scene,—the long retiring vista on either hand, purple in the far distance; and remember that that vast rectilinear hollow forms but one of the plough furrows of my illustration,—they will see that that with which I am in reality dealing is the sublime of nature, and that even the details of my subject, rightly appreciated, are not suited to lower our conceptions of the wonderful workings of old of Him, who, by processes which science is but now aspiring to comprehend, 'gathered the waters together into one place, that the dry land might appear,' and laid the deep-seated foundation of the mighty hills.

Let us now pass on to the Oolite proper, and its base the Lias, as we find them developed in Scotland. They form but a comparatively small portion of the surface of the country,—not much more, it has been estimated, than sixty square miles; nor can I refer definitely to any marked peculiarity of scenery in the districts in which they occur. The Oolites of Sutherland extend westwards and southwards from the Ord Hill of Caithness to the village of Golspie, a distance of about sixteen miles; and form, under the rugged line of hills against whose flanks they recline, a green narrow strip of low country, that, where not too deeply covered up by débris of the Primary rocks, transported from the interior during the Pleistocene period, is, for its extent, of great agricultural value, and bears on its cultured surface the rich fields and extensive woods of Dunrobin, the stately castle of the old Earls of Sutherland. Further to the west and south, along the eastern shores of Cromarty and Ross, detached patches of the Lias occur, as at Shandwick, at the Northern Sutor of Cromarty, at the Southern Sutor, and at the Hill of Eathie,—each patch occurring directly opposite, and leaning against, one of the upheaved hills, which, as I have already said, were undoubtedly the agents in raising and bringing it to the surface. The Lias and Oolite also appear on the southern side of the Moray Firth, in the counties of Moray and Banff, but merely as outliers of very limited extent, and sorely broken up or ground down by the denuding Pleistocene agencies. On the western coast of Scotland the Lias may be seen on the mainland at Applecross, and on the sides of Loch Aline, opposite the Sound of Mull ; while in the inner Hebrides, it forms, with the Oolite, though greatly overflown by trap, the base of the larger part of the island of Mull, of two of the Small Isles, Eigg and Muck, of Raasay and Scalpa, and of large tracts of the eastern and northern half of Skye. At Broadford, in the latter island, the Lias forms the whole of the rich level islet of Pabba, which, lying as at anchor in its quiet bay, reminds one, from its prevailing colour and form, of one of the low, green steamboats of the Clyde. Opposite Pabba, the Liassic deposit sweeps across the mainland of Skye from sea to sea, along a flat valley some two or three miles wide ; but while the minute Liassic islet resembles, from the softness of its outline, an islet of England set down in a hill-enclosed bay of the Scottish Highlands, there is nothing English in the scenic character of the Liassic valley. It is a brown and sombre expanse of marsh and moor, studded by blue dreary lochans, interesting, however, to the botanist as habitats of the rare *Eriocaulon septangulare*. The waste is haunted, too, say the Highlanders, by Ludag, a malignant goblin, not more known elsewhere in Europe than the rare plant that in the last age used to be seen at dusk hopping with immense hops on its one leg,—for, unlike every other denizen of the supernatural world, it is not furnished with two,—and that, enveloped in rags, and with fierce misery in its hollow eye, has dealt heavy blows, it is said, on the cheeks of benighted travellers. Certainly a more appropriate spectre could scarce be summoned to walk at nights over the entombed remains of the old monsters of the Lias than one-legged Ludag, the goblin of the wastes of Broadford. Such, in brief, is a summary of our Oolitic deposits. They occupy, as I have said, but a small portion of the surface of Scotland ; and, though coal has occasionally been wrought in them, and though they furnish in several localities supplies of lime and of building stone, their economic importance is comparatively small. But a well-filled volume,—the life-long work of some laborious chronicler,—may have *no* economic importance in the lower and humbler sense, and may yet form a valuable record of bygone transactions and events suited to delight and instruct throughout all generations. And it is thus with the Oolitic deposits of Scotland. Their innumerable strata, closely written 'within and without' in a language in which every character is an organism, form the leaves of a record in which many of the marvellous existences that flourished during what are geologically the middle ages of our country's history are well and wonderfully preserved. Instead of dissipating your attention by describing at length the fossils of its various deposits, I shall attempt giving you a general idea of the whole under the ordinary division of animal and vegetable, as they have come to my knowledge during the researches of at least thirty years.

In one of its features the Oolitic flora of what is now Scotland must have resembled its flora in the present, or rather in the past age, ere our native pine-woods had yielded to the axe. Trees of the fir or pine division of the Coniferæ, many of them of slow growth and large size, must have formed huge forests in a province of the land of the Oolite which extended from what is now the island of Mull to the Ord Hill of Caithness. The Scuir of Eigg, a sub-aërial mole of columnar pitchstone, four hundred feet in height, and perched on the ridge of a tall hill, rests on the remains of a prostrated forest, as some of our submarine moles rest on foundations of piles. And of this forest all the trees seem to have consisted of one species,—a conifer of the Oolite now known to the fossil botanist as the *Pinites Eiggensis*, or Eigg pine. Branches and portions of the trunks of a similar pine are not unfrequent in the Lias of Eathie and Ross; and in shale-beds of the Lower Oolite in the neighbourhood of Helmsdale there occur in abundance fossil trunks and branches, mingled with cones and the narrow spiky leaflets characteristic of the family. I have reckoned in the transverse section of a Helmsdale pine-trunk about two feet in diameter, more than a hundred annual rings. And from the rings and roots of some of the others, its contemporaries, I found that curious insight might be derived respecting the state and condition of vegetable life in the old Scotch woods of the Oolite. In the first place, the annual rings themselves told me, when exposed to transmitted light in the microscope, that the winters of that time gave vegetation as decided a check as our winters now. The tender woody cells were first dwarfed and thickened in their formation by the strengthening of the autumnal cold, and then for a season they ceased to form altogether. But then the spring came, and over the hard concentric line drawn by the chill hand of winter they began to form themselves anew in full-sized luxuriance; and thus, year after year, and for century after century, the process went on. Some of these ancient pine-trees grew in rich sheltered hollows, and acquired bulk so rapidly, that they increased their diameter eight and a half inches in twenty years; others grew so slowly, that they increased their diameter only two and a half inches in forty years. And it is a curious circumstance, that in both those of slower and of more rapid growth we find alternating groups of broader and narrower annual rings, indicating apparently groups of better and worse seasons. Lord Bacon remarks in one of his Essays,—the Essay on the Vicissitude of Things,—that it was a circumstance first observed in the Low Countries (the provinces of the Netherlands), that there were certain meteorological cycles of seasons,—groups of warmer and groups of colder summers, and of more temperate and of less temperate winters,—which periodically came round again. And we have seen not very successful attempts made in our own times to measure these cycles, and reduce them to a formula, from which the nature of the coming seasons might be determined beforehand. But there can be little doubt,—whatever the cause or the order of their occurrence,—that alternations of groups of colder and warmer, better and worse seasons, do occur; and it seems more than probable that, in obedience to some occult law, as little understood in the present age as when its operations were first detected in the Netherlands, Scotland had in the times of the Oolite, as certainly as now, its alternating groups of chill and of genial summers, and of temperate and severe winters. And the well-marked rings of its fossil Coniferæ remain to attest the fact. We can even determine the kind of soil into which a certain proportion of these ancient pines struck root. It was extremely shallow in some localities, and lay over a hard bottom. We find that some of the fossil stumps shot out their roots horizontally immediately as they entered the earth, and sent down no vertical prolongations of the trunk into the subsoil,—an arrangement still common among the roots of trees planted on a shallow stratum of soil resting on a hard bottom. Further, we are still able to ascertain that the hard bottom that underlay the soil in which some of the Oolitic pines of Helmsdale grew was composed of Old Red flagstone, identical in its mineral composition and organic remains with what is now known as Caithness flag.

But let us trace the history of a single pine-tree of the Oolite, as indicated by its petrified remains. This gnarled and twisted trunk once anchored its roots amid the crannies of a precipice of dark-grey sandstone, that rose over some nameless stream of the Oolite, in what is now the north of Scotland. The rock, which, notwithstanding its dingy colour, was a deposit of the Lower Old Red Sandstone, formed a member of the fish-beds of that system,—beds that were charged then, as now, with numerous fossils, as strange and obsolete in the creation of the Oolite as in the creation which at present exists. It was a firm, indestructible stone, covered by a thin, barren soil; and the twisted rootlets of the pine, rejected and thrown backwards from its more solid planes, had to penetrate into its narrow fissures for a straitened and meagre subsistence. The tree grew but slowly: in considerably more than half a century it had attained to a diameter of little more than ten inches a foot over the soil; and its bent and twisted form gave evidence of the life of hardship to which it was exposed. It was, in truth, a picturesque rag of a tree, that for the first few feet twisted itself round like an overborne wrestler struggling to escape from under his enemy, and then struck out at an abrupt angle, and stretched itself like a bent arm over the stream. It must have resembled, on its bald eminence, that pine-tree of a later time described by Scott, that high above 'ash and oak,'

> 'Cast anchor in the rifted rock,
> And o'er the giddy chasm hung
> His shatter'd trunk, and frequent flung,
> Where seem'd the cliffs to meet on high,
> His boughs athwart the narrow'd sky.'

The seasons passed over it: every opening spring gave its fringe of tenderer green to its spiky foliage, and every returning autumn saw it shed its cones into the stream below. Many a delicate fern sprang up and decayed around its gnarled and fantastic root, single-leaved and simple of form, like the *Scolopendria* of our caverns and rock recesses, or fretted into many a slim pinnate leaflet, like the minute *maiden-hair* or the graceful *lady-fern*. Flying reptiles have perched amid its boughs; the light-winged dragon-fly has darted on wings of gauze through the openings of its lesser twigs; the tortoise and the lizard have hybernated during the chills of winter amid the hollows of its roots; for many years it formed one of the minor features in a wild picturesque scene, on which human eye never looked; and at length, touched by decay, its upper branches began to wither and bleach white in the winds of heaven; when shaken by a sudden hurricane that came roaring down the ravine, the mass of rock in which it had been anchored at once gave way, and, bearing fast jammed among its roots a fragment of the mass which we still find there, and from which we read a portion of its story, it was precipitated into the foaming torrent. Dancing on the eddies, or lingering amid the pools, or shooting, arrow-like, adown the rapids, it at length finds its way to the sea; and after sailing over beds of massive coral,—the ponderous *Isastrea* and more delicate *Thamnastrea*,—and after disturbing the Enaliosaur and Belemnite in their deep-green haunts, it sinks, saturated with water, into a bed of arenaceous mud, to make its appearance, after long ages, in the world of man,—a marble mummy of the old Oolite forests,—and to be curiously interrogated regarding its character and history.

The pines of our Scotch Oolite—some of them, as I have shown, or rather as my specimens show, of exceedingly slow growth—are suggestive of a temperate, if not severe climate. The family of their contemporaries, however, to which I must next refer as not less characteristic of the flora of this ancient time than the coniferæ themselves, is now to be found in a state of nature in only the warmer regions of the earth, and can be studied in this part of the world only in our conservatories and greenhouses. It is known to the botanist as the Cycadaceous family; and at least two of its genera, Cycas and Zamia, we find well represented in the Oolitic deposits of Scotland. In the Zamia, a cylindrical, squat, scale-covered pedestal is fringed along its upper edge by a ring of long pinnate leaves, that radiate outwards like the spokes of a wheel from the nave; and, placed on the centre of the pedestal, there is, when the plant is in fruit, a handsome cone. The *tout ensemble* is as if a pine-apple, with the pot in which it grew, and with its leaves arranged like a ruff round its stem, formed altogether but one plant. The Cycas is usually taller than

Zamia; the leaves also, of the compound pinnate character, are smaller and more bushy; and it resembles, as a whole, a decapitated palm, with a coronal of fern bound atop, as if to conceal the mutilation. With these Cycadaceæ there flourished in the marshes of the period plants of a family still widely spread over the various climatal zones, but which now attain to any considerable size only within the tropics. I refer to the Equisetaceæ, or horse-tail family,— slim, cone-crowned plants, fringed with green verticillate leaves, or branches rather, and which in this country are rarely thicker than a quill, or rarely exceed eighteen inches in height, but which have been found in the intertropical swamps of South America fifteen feet high, and three inches in circumference at the lower part of the stem. In the Oolite of Scotland, a well-marked, long-extinct species, the *Equisetum columnare* must have attained, judging from the thickness of the stem, which is sometimes fully three inches in diameter, to at least thrice the size of its tropical congeners. As shown by its remains, which occur in the lignite shales of Brora, it must have been a plant of considerable elegance of form, encircled at each joint at some of the specimens by torus-like mouldings grooved crosswise, traversed in the spaces between by longitudinal markings, delicately punctulated, and gracefully feathered from root to pointed top by its verticillate garlands of spiky leaves. The Lycopodiaceæ or club-moss family, existing in rather massier and more arboraceous forms than now, though reduced in a greatly more than equal degree from their gigantic congeners of the Coal Measures, were also abundant (as shown by the rocks of Helmsdale) in the Oolitic flora of Scotland; and with these there mingled various genera, consisting of numerous species of well-marked ferns. Ferns, indeed, so far as we yet know, may be regarded as forming the base, and pines the apex, of the terrestrial Oolitic flora; and between these two extremes most of its other productions seem to have ranged. The Cycadaceæ possess certain characters which belong to both : they are, if I may so speak, fern-pines, with, in some instances, a peculiarity of aspect which seems also to ally them to the palms. Again, the Lycopodiaceæ, intermediate between the mosses and the ferns, may be described as fern-mosses, with a peculiarity of aspect in some of the Oolitic species that seems to ally them to the pines. And the Equisetaceæ belong to at least the same sub-class as the ferns,—the Acrogens. The Palmæ, as shown by the English deposits, were also present in the Oolitic flora : nor is it probable that a species of vegetation which the old Yorkshire of the Oolite possessed, the old Scotland of the Oolite should have wanted; though I have not yet succeeded in finding the remains of palms in any of our Scotch deposits.

The animal productions of our country during this early period were divided, like those of the present time, into the four great Cuvierian divisions, all of which we still find in a fossil state in our rocks. Corals akin to the tropical forms, —some of them of great size,—with star-fishes and sea-eggs, represent the radiata; a fossil lobster which occurs in the Lias of Cromarty somewhat meagrely represents the articulata. The shelled mollusca we find very largely represented in almost all their classes and families, from the high Cephalopods to the low Brachipods; and in this division the peculiar character of the Oolitic system is more strongly impressed than even on its flora. Its corals, though many of them of great size, as I have just said, and of elegant form, might almost pass for those of the intertropical seas of the present day; nor are its crustacea and insects, even where best preserved, as in the Oolites of England, of a character widely different from those which still exist. But by much the larger part of its mollusca bear the stamp of a fashion that has perished. It is chiefly, however, in its molluscas of the first class,—the Cephalopods,—creatures of a high standing in their division, and represented in the present day by the nautilus and the cuttle-fish, that we recognise in its fullest extent this extinct peculiarity of type and form. Its Brachipods, chiefly terebratulæ, not unfrequent in the Sutherland Oolites, and in the Lias of Cromarty and Skye,—its periwinkles, whelks, aviculæ, pinnæ, pectens, oysters, and mussels, few of them wanting in any of our Scotch Liassic or Oolitic deposits, and many of them very abundant, though all specifically extinct, present us, though with a large admixture of strange and exotic forms, with many other forms with which, generically at least, we are familiar. But among the Cephalopods all is strange and unwonted; and their vast numbers—greater at this period of the world's history than in any former or any after time—have the effect of imparting their own unfamiliar character to the whole molluscan group of the Oolite. I need but refer to two families of these,—the Belemnite family and the family of the Ammonites; both of them so remarkable, that they attracted in their rocks the notice of the untaught inhabitants of both England and Scotland, and excited their imagination to the point at which myths and fables are produced, long ere Geology existed as a name or was known as a science. The Belemnites are the old thunder-bolts of the north of Scotland, that, in virtue of their supposed descent from heaven, were deemed all potent in certain cases of bewitchment and the evil-eye; and the Ammonites are those charmed snakes of the mediæval legend,

> 'That each one
> Was changed into a coil of stone,
> When holy Hilda prayed.'

The exact affinities of the Belemnite family have formed a subject of controversy of late years among our highest authorities,—men such as Professor Owen taking up one side, and men such as Dr. Mantell the other. But there can be little doubt that it more nearly approached to our existing cuttle-fishes than to any other living animals; while there is no question that its contemporary the Ammonite is now most nearly represented, though of course only approximately, by the nautilus. The Belemnite existed in some of its species throughout all the formations of the great Secondary division, but neither during those of the Palæozoic nor yet of the Tertiary divisions; the Ammonite, on the other hand, though in an extreme and aberrant form, preceded it by several formations, but became extinct at the same time—neither Ammonite nor Belemnite outliving the deposition of the Chalk.

The first great division of the animal kingdom, the vertebrata, was represented in Scotland during the Oolitic period by fishes and reptiles. Its fishes seem to have been restricted to two orders,—that placoid order to which the existing sharks belong, and that ganoid order, now wellnigh worn out in creation, to which the *Lepidosteus* of the North American lakes and rivers belongs, and to which I incidentally referred in connexion with the Lepidotus of the Weald. I have found in the island of Eigg beds of a limestone composed almost entirely of fossil shells, which were strewed over with the teeth of an extinct genus of sharks, the Hybodonts; and I have seen the dorsal spines of the same placoid division occasionally occurring among the Oolites of Sutherland and the Lias of Eathie. And scales, cerebral plates, and in some instances considerable portions of individuals of the ganoidal species, glittering in the enamel to which they owe their name, occur in all the Oolitic deposits of Scotland. Of our Scottish reptiles of the Oolite we have still a good deal to learn. I was fortunate enough in 1844 to find in a deposit of Eigg, and again at Helmsdale, in 1849, the remains of several of its more characteristic Enaliosaurs, or bepaddled reptiles of the sea; at Helmsdale I found vertebral joints of the Ichthyosaurus in a conglomerate lower in the Oolite; and in Eigg, in a stratum composed of littoral univalves, vertebral joints, phalanges, and portions of the humerus and of the pelvic arch of Plesiosaurus, together with the limb-bones of crocodileans, and fragments of the carapace of a tortoise. Previous, however, to even the earlier date of my discoveries, the tooth of a Saurian had been found in the Sutherlandshire Oolite by Mr. (now Sir Roderick) Murchison, and the limb-bone of a Chelonian with a sauroid vertebra, in the outlier of the Morayshire Weald at Linksfield. My collection, however, though still very inadequate in this department, contains, in quantity at least, and, I am disposed to think, in variety also, some eight or ten times more of the reptilian remains of Scotland, during the Secondary ages, than all the other collections of the kingdom. They at least serve to demonstrate that the Oolitic period in what is now our country, was, as in England and on the Continent, a period of huge and monstrous reptiles,—that the bepaddled Enaliosaurs, the strange reptilian predecessors of the Cetacea, haunted our seas in at least two of their generic forms,— that of the Ichthyosaur and that of the Plesiosaur ; that our rivers were frequented by formidable crocodiles ; and that tortoises of various perished species lived in our lakes and marshes, or, according to their natures, disported on the drier grounds. Nor is it probable that the other reptilian monsters of the time, the contemporaries of these creatures in England, would have been wanting here. We may safely infer that flocks of Pterodactyles,—reptiles mounted on bat-like wings, and as wild and monstrous in aspect and proportion as romancer of the olden time ever feigned,— fluttered through the tall pine-forests, or perched on the cycadeæ and the tree-ferns ; that the colossal Iguanodon and gigantic Hylæosaurus browsed on the succulent equisetaceæ of the low meadows ; that the minute Amphitherium, an insectivorous mammal of the period, lodged among the ferns on the drier grounds, where extinct grasshoppers chirped throughout the long bright summer, and antique coleoptera burrowed in the sand; and that far off at sea there were moments when the sun gleamed bright on the polished sides of the enormous Cetiosaurus, as it rose from the bottom to breathe. But I must close this part of my subject,—the Scottish flora and fauna of the Oolite, —on which my narrow limits permit me, as you see, to touch at merely a few salient points,—with two brief remarks :—*First*, So rich was its flora, that its remains formed on the east coast of Sutherland a coal, or rather lignite field, so considerable that it was wrought for greatly more than a century,—at one time to such effect, that during the twelve years which intervened between 1814 and 1826, no fewer than seventy thousand tons of coal were extracted from one pit. *Second*, The strange union which we find in the same beds of trees that seem to have languished under chill and severe skies, with plants, corals, and shells of a tropical or semi-tropical character, need not be regarded as charged with aught like conflicting evidence respecting the climatal conditions of the time. Climate has its zones marked out as definitely by thousands of feet on our hill-sides as by degrees of latitude on the surface of the globe; and if the Scotland of the Oolitic period was, as is probable, a mountainous country traversed by rivers, productions of an intertropical, and of even a semi-arctic, character, may have been not only produced within less than a day's journey of each other, but their remains may have been mingled by land-floods, as we find the huge corals of Helmsdale blent with its slow-growing pines, among the débris of some littoral bed. The poet's exquisite description of Lebanon suggests, I am disposed to think, the true reading of the enigma :—

> ' Like a glory the broad sun
> Hangs over sainted Lebanon,
>   Whose head in wintry grandeur towers,
>   And whitens with eternal sleet ;

While summer, in a vale of flowers,
Is sleeping rosy at his feet.'

The mere lists of the botanist and zoologist are in themselves repulsive and un-ideaed; and yet the existences which their arbitary signs represent are the vital marvels of creation,—the noble forests, fair shrubs, and delicate flowers, and the many-featured denizens of the animal world, so various in their forms, motions, and colours, and so wondrous in their structure and their instincts. I have been presenting you this evening with little else than a dry list of the Scottish productions of the Wealden and Oolitic ages,—a list necessarily imperfect, and all the more unsuggestive from the circumstance that, as myriads of ages had elapsed between the extinction of the races and families which its signs represent, and their first application as signs, so these signs, in their character as vocables, belong to languages as dead as the organisms themselves. The organisms were dead and buried, and converted into lignite or stone, long ages ere there was language enough in the world to furnish them with names; and now the dead has been employed to designate the dead,—dead languages to designate the remains of dead creations. Could we but see the productions of our country as they once really existed,—could we travel backwards into the vanished past, as we can descend into the strata that contain their remains, and walk out into the woods, or along the sea-shores of old Oolitic Scotland, we should be greeted by a succession of marvels strange beyond even the conceptions of the poet, or at least only equalled by the creations of him who, in his adventurous song, sent forth the Lady Una to wander over a fairy land of dreary wolds and trackless forests, whose caverns were haunts of dragons and satyrs, and its hills the abodes

'Of dreadful beasts, that, when they drew to hande,
Half-flying and half-floating, in their haste,
Did with their largeness measure o'er much lande,

And made wide shadow under bulksome waist,
As mountain doth the valley overcaste ;
And trailing scaly tails did rear afore
Bodies all monstrous, horribill, and vaste.'

Let us, however, ere we part for the evening, adventure a short walk into the wilds of the Oolite, in that portion of space now occupied on the surface of the globe by the north-eastern hills of Sutherland, where they abut on the precipitous Ord.

We stand on an elevated wood-covered ridge, that on the one hand overlooks the blue sea, and descends on the other towards a broad river, beyond which there spreads a wide expanse of a mountainous, forest-covered country. The higher and more distant hills are dark with pines; and, save that the sun, already low in the sky, is flinging athwart them his yellow light, and gilding, high over shaded dells and the deeper valleys, cliff, and copse, and bare mossy summit, the general colouring of the background would be blue and cold. But the ray falls bright and warm on the rich vegetation around us,—tree ferns, and tall club-mosses, and graceful palms, and the strangely-proportioned cycadaceæ, whose leaves seem fronds of the bracken fixed upon decapitated stumps; and along the banks of the river we see tall, intensely green hedges of the feathered equisetaceæ. Brown cones and withered spiky leaves strew the ground; and scarce a hundred yards away there is a noble Araucarian, that raises, sphere-like, its proud head more than a hundred feet over its fellows, and whose trunk, bedewed with odoriferous balsam, glistens to the sun. The calm stillness of the air makes itself faintly audible in the drowsy hum of insects; there is a gorgeous light-poised dragon-fly darting hither and thither through the minuter gnat-like groups : it settles for a moment on one of the lesser ferns, and a small insectivorous creature, scarce larger than a rat, issues noiselessly from its hole, and creeps stealthily towards it. But there is the whirr of wings heard overhead, and lo! a monster descends, and the little mammal starts back into its hole. 'Tis a winged dragon of the Oolite, a carnivorous reptile, keen of eye and sharp of tooth, and that to the head and jaws of the crocodile adds the neck of a bird, the tail of an ordinary mammal, and that floats through the air on leathern wings resembling those of the great vampire bat. We have seen, in the minute, rat-like creature, one of the two known mammals of this vast land of the Oolite,—the insect-eating *Amphitherium;* and in the flying reptile, one of its strangely organized *Pterodactyls.*

But hark! what sounds are these? Tramp, tramp, tramp, —crash, crash. Tree-fern and club-moss, cycas and zamia, yield to the force and momentum of some immense reptile, and the colossal *Iguanodon* breaks through. He is tall as the tallest elephant, but from tail to snout greatly more than twice as long; bears, like the rhinoceros, a short horn on his snout; and has his jaws thickly implanted with saw-like teeth. But, though formidable from his great weight and strength, he possesses the comparative inoffensiveness of the herbivorous animals; and, with no desire to attack, and no necessity to defend, he moves slowly onward, deliberately munching, as he passes, the succulent stems of the cycadaceæ. The sun is fast sinking, and, as the light thickens, the reaches of the neighbouring river display their frequent dimples, and ever and anon long scaly backs are raised over its surface. Its numerous crocodileans are astir; and now they quit the stream, and we see its thick hedge-like lines of equisetaceæ open and again close, as they rustle through, to scour, in quest of prey, the dank meadows that line its banks. There are tortoises that will this evening find their protecting armour of carapace and plastron all too weak, and close their long lives of centuries. And now we saunter downwards to the shore, and see the ground-swell breaking white in the calm against ridges of coral scarce less white. The shores are strewed with shells of pearl,—the whorled Ammonite and the Nautilus; and amid the gleam of ganoidal scales, reflected from the green depths beyond, we may see the phosphoric trail of the Belemnite, and its path is over shells of strange form and name,—the sedentary Gryphæa, the Perna, and the Plagiostoma.

But lo! yet another monster. A snake-like form, surmounted by a crocodilean head, rises high out of the water within yonder coral ledge, and the fiery sinister eyes peer inquiringly round, as if in quest of prey. The body is but dimly seen; but it is short and bulky compared with the swan-like neck, and mounted on paddles instead of limbs; so that the entire creature, wholly unlike anything which now exists, has been likened to a boa-constrictor threaded through the body of a turtle. We have looked upon the *Plesiosaurus*. And now outside the ledge there is a huge crocodilean head raised; and a monstrous eye, huger than that of any other living creature,—for it measures a full foot across,—glares upon the slimmer and less powerful reptile, and in an instant the long neck and small head disappear. That monster of the immense eye,—an eye so constructed that its focus can be altered at will, and made to comprise either near or distant objects, and the organ itself adapted either to examine microscopically or to explore as a telescope,—is another bepaddled reptile of the sea, the *Ichthyosaurus* or fish-lizard. But the night comes on, and the shadows of the woods and rocks deepen: there are uncouth sounds along the beach and in the forest; and new monsters of yet stranger shape are dimly discovered moving amid the uncertain gloom. Reptiles, reptiles, reptiles,—flying, swimming, waddling, walking;—the age is that of the cold-blooded, ungenial reptile; and, save in the dwarf and inferior forms of the marsupials and insectivora, not one of the honest mammals has yet appeared. And now the moon rises in clouded majesty; and now her red wake brightens in one long strip the dark sea; and we may mark where the Cetiosaurus, a sort of reptilian whale, comes into view as it crosses the lighted tract, and is straightway lost in the gloom. But the night grows dangerous, and these monster-haunted woods were not planted for man. Let us return then to the safer and better furnished world of the present time, and to our secure and quiet homes.

LECTURE FIFTH.

The Lias of the Hill of Eathie—The Beauty of its Shores—Its Deposits, how formed—Their Animal Organisms indicative of successive Platforms of Existences—The Laws of Generation and of Death—The Triassic System—Its Economic and Geographic Importance—Animal Footprints, but no Fossil Organisms, found in it—The Science of *Ichnology* originated in this fact—Illustrated by the appearance of the Compensation Pond, near Edinburgh, in 1842—The Phenomena indicated by the Footprints in the Triassic System—The Triassic and Permian Systems once regarded as one, under the name of the New Red Sandstone—The Coal Measures in Scotland next in order of Succession to the Triassic System—Differences in the Organisms of the two Systems—Extent of the Coal Measures of Scotland—Their Scenic Peculiarities—Ancient Flora of the Carboniferous Period—Its Fauna—Its Reptiles and Reptile Fishes—The other Organisms of the Period—Great Depth of the System—The Processes by which during countless Ages it had been formed.

THE Lias forms, as I have already had occasion to remark, the base of the great Oolitic system. I dealt in my last address with the productions, vegetable and animal, of those long ages of the world's history which the various deposits of this system represent, and attempted a restoration of some of its more striking scenes, as they must have existed of old in what is now Scotland. But in glancing once more at the Lias, we must pass from the living to the dead, from the vital myriads that once were, to the cemetery that contains their remains. I shall select as my example a single Liassic deposit of Scotland, but in several respects one of the most remarkable,—that of Eathie, on the shores of the Moray Firth, about four miles from the town of Cromarty. And in visiting it in its character as a great burial-ground,—the final resting-place, not only of perished individuals, but also of extinct tribes and races, and in scanning its strangely sculptured monuments, roughened with hieroglyphics, to which living nature furnishes the key, we may perhaps be permitted to indulge in some of those reflections which so naturally suggest themselves in solitary churchyards, or among the tombs of the ancient dead.

The hill of Eathie is a picturesque eminence of granitic gneiss, largely mixed with beds of hornblende schist, which extends, in a long precipitous ridge, some five or six hundred feet in height, along the northern side of the Moray Firth, and forms one of a primary chain of hills, which, in their upheaval, uptilted deposits of the Lias and Oolite. The deposit which the hill of Eathie disturbed is exclusively a Liassic one: the upturned edge of the base of the formation rests against the bottom of the hill; and we may trace the edges of its various upper deposits for several hundred feet outwards, bed above bed, until, apparently near the top of the formation, we lose them in the sea. There is a wild beauty on the shores of Eathie. A selvage of comparatively level ground, that occupies the space between the rocky beach and an inflection of the hill, seems embosomed in solitude; the naked scaurs and furze-covered slopes, where the fox and the badger breed, interpose their dizzy fence between it and the inhabited portions of the country above; while the rough unfrequented shore and wide-spreading sea form the secluding barriers below. The only human dwellings visible are the minute specks of white that look out in the sunshine from the dim and diluted blue of the opposite coast; and we may see the lonely firth broadening and widening as it recedes from the eye, and opens to the ocean in a direction so uninterrupted by land, that the waves, which, when the wind blows from the keen north, first begin to break on the distant headlands, and then come running up the coast, like white coursers, may have heaved their first undulating movements under the polar ice. The scene seems such a one as the anchorite might choose to wear out life in, far from the society of fellow-man; and we actually find, in exploring its bosky thickets of wild rose and sloe-thorn, that some anchorite of the olden time *did* make choice of it. A grey shapeless hillock of lichened stone, shaded by luxuriant tufts of fern, still bears the name of the old chapel; and an adjacent spring, on whose overhanging sprays of ivy we may occasionally detect minute tags of linen and woollen cloth,—the offerings of a long-derived superstition, not quite extinct in the district,—is still known as the Saint's Well. But who the anchorite was, tradition has long since forgot; and it was only last year that I succeeded in recovering the name of the saint from an old man, whose father had been a farmer on the land considerably more than a hundred years before. The chapel and spring had been dedicated, he said, to St. Kennat,—a name which we need scarce look for in the Romish Calendar, but which designated, it is probable, one of our old Culdee saints.

The various beds of the Eathie deposit,—all save the lowest, which consists of a blue adhesive clay,—are composed of a dark, finely laminated shale; and, varying in thickness from thirty feet to thirty yards, they are curiously separated from each other by bands of fossiliferous limestone. And so impalpable a substance are these shales, that, when subjected to calcination, which is necessary to extract the bitumen with which they are charged, and which gives them toughness and coherency, they resolve into a powder, used occasionally, from its extreme fineness, in the cleaning of polished brass and copper. They were laid down, it is probable, in circumstances similar to those in which, as described by the late Captain Basil Hall, extensive deposits are now taking place in the Yellow Sea of China. 'At sunset,' says Captain Hall, in the narrative of his voyage to Loo-Choo, 'no land could be perceived from the mast-head, although we were in less than five fathoms water. And before the day broke next morning, the tide had fallen a whole fathom, which brought the ship's bottom within three feet of the ground. It was soon afterwards discovered that she was actually sailing along with her keel in the mud, which was sufficiently indicated by a long yellow train in our wake. Some inconvenience was caused by this extreme shallowness, as it retarded our headway, and affected the steering; but there was in reality not much danger, as it was ascertained, by forcing long poles into the ground, that for many fathoms below the surface on which the sounding lead rested, and from which level the depth of water is estimated, the bottom consisted of nothing but mud formed of an impalpable powder, without the least particle of sand or gravel.' The Liassic deposit of Eathie must have been of slow deposition. It consists of laminæ as thin as sheets of pasteboard, which, of course, shows that there was but little deposited at a time, and pauses between each deposit. And, though a soft muddy surface could have been of itself no proper habitat for the sedentary animals,—serpulæ, oysters, gryphites, and terebratulæ,—we find further, that they did, notwithstanding, find footing upon it, by attaching themselves to the dead shells of such of the sailing or swimming molluscs, Ammonites and Belemnites, as died over it, and left upon it their remains; from which we infer that the pauses must have been very protracted, seeing that they gave time sufficient for the Terebratulæ,—shells that never moved from the place in which they were originally fixed,—to grow up to maturity. The thin leaves of these Liassic volumes must have been slowly formed and deliberately written; for as a series of volumes, reclining against a granite pedestal in the geologic library of nature, I used to find pleasure in regarding them. The limestone bands, curiously marbled with lignite, ichthyolite, and shell, formed the stiff boarding; and the thin pasteboard-like laminæ between,—tens and hundreds of thousands in number in even the slimmer volumes,—composed the closely written leaves. For never did characters or figures lie closer in a page than the organisms on the surfaces of these leaf-like laminæ. Permit me to present you from my note-book with a few readings taken during a single visit from these strange pages.

We insinuate our lever into a fissure of the shale, and turn up a portion of one of the laminæ, whose surface had last seen the light when existing as part of the bottom of the old Liassic sea, when more than half the formation had still to be deposited. Is it not one of the prints of Sowerby's *Mineral Conchology* that has opened up to us? Nay, the shells lie too thickly for that, and there are too many repetitions of organisms of the same species. The drawing, too, is finer, and the shading seems produced rather by such a degree of relief in the figures as may be seen in those of an embossed card, than by any arrangement of lighter and darker colour. And yet the general tone of the colouring, though dimmed by the action of untold centuries, is still very striking. The ground of the tablet is of a deep black, while the colours stand out in various shades, from opaque to silvery white, and from silvery white to deep grey. *There*, for instance, is a group of large Ammonites, as if drawn in white chalk; *there*, a cluster of minute bivalves resembling Pectens, each of which bears its thin film of silvery nacre; *there*, a gracefully formed Lima in deep neutral tint; while, lying athwart the page, like the dark hawthorn leaf in Bewick's well-known vignette, there are two slim sword-shaped leaves coloured in deep umber. We lay open a portion of another page. The centre is occupied by a large Myacites, still bearing a warm tint of yellowish brown, and which must have been an exceedingly brilliant shell in its day; there is a Modiola, a smaller shell, but similar in tint, though not quite so bright, lying a few inches away, with an assemblage of dark grey Gryphites of considerable size on the one side, and on the other a fleet of minute Terebratulæ, that had been borne down and covered up by some fresh deposit from above, when riding at their anchors. We turn over yet another page. It is occupied exclusively by Ammonites of various sizes, but all of one species, as if a whole argosy, old and young, convoyés and convoyed, had been wrecked at once, and sent disabled and dead to the bottom. And here we open yet another page more. It bears a set of extremely slender Belemnites. They lie along and athwart, and in every possible angle, like a heap of boarding-pikes thrown carelessly down on a vessel's deck on the surrender of the crew. Here, too, is an assemblage of bright black plates, that shine like pieces of japan work, the cerebral plates of some fish of the ganoid order; and here an immense accumulation of minute glittering scales of a circular form. We apply the microscope, and find every little interstice in the page covered with organisms. And leaf after leaf, for tens and hundreds of feet together, repeats the same strange story. The great Alexandrian library, with its unsummed tomes of ancient literature, the accumulation of long ages, was but a poor and meagre collection, scarce less puny in bulk than recent in date, when compared with this vast and wondrous library of the Scotch Lias.

Now, this Eathie deposit is a crowded burying-ground, greatly more charged with remains of the dead, and more thoroughly saturated with what was once animal matter, than ever yet was city burying-ground in its most unsanitary state. Every limestone band or nodule yields, when struck by the hammer, the heavy fetid odour of corruption and decay; and so charged is the laminated shale with an animal-derived bitumen, that it flames in the fire as if it had been steeped in oil, and yields a carburetted hydrogen gas scarce less abundantly than some of our coals of vegetable origin. The fact of the existence, throughout all the geological ages, of the great law of death, is a fact which must often press upon the geologist. Almost all the materials of his history he derives from cenotaphs and catacombs. He finds no inconsiderable portion of the earth's crust composed of the remains of its ancient inhabitants,—not of dead individuals merely, but also of dead species, dead genera, nay, of even dead creations; and here, where the individual dead lie as thickly on the surface of each of many thousand layers as leaves along the forest glades in autumn,—here, where all the species and many of the genera are dead, nay, where the whole creation represented by its multitudinous organisms is dead,—the great problem which this law of death presents comes upon the explorer in its most palpable and urgent form. The noble verses of James Montgomery, somewhat exaggerative in their character when addressed to a molehill, become as remarkable for their sober propriety as for their beauty when employed here :—

'Tell me, thou dust beneath my feet,
　Thou dust that once hadst breath,—
Tell me how many mortals meet
　In this small hill of death.

By wafting winds and flooding rains,
　From ocean, earth, and sky,
Collected, here the frail remains
　Of slumbering millions lie.

The mole that scoops, with curious toil,
　Her subterranean bed,
Thinks not she ploughs so rich a soil,
　And mines among the dead.

But oh! where'er she turns the ground,
　My kindred earth I see;
Once every atom of this mound
　Lived, breathed, and felt like me.

Like me, these elder-born of clay
　Enjoyed the cheerful light,
Bore the brief burden of a day,
　And went to rest at night.

Methinks this dust yet heaves with breath,
　Ten thousand pulses beat :
Tell me, in this small hill of death
　How many mortals meet.'

What does this inexorable law of death mean, or on what principle does it depend? In our own species it has a moral significancy,—'Death reigned from Adam;' and, though a pardonable mistake, no longer insisted on by at least theologians of the higher class, the same moral character, as a reflex influence, has been made to attach to it in its inevitable connexion with the inferior animals. But in them it seems to have no moral significancy. Bacon makes a shrewd distinction, in one of his Essays, between 'death as the wages of sin,' and death as 'a tribute due to nature;' and we can now fully appreciate the value of the distinction. For we now know that while as the wages of sin it has reigned from but the fall of Adam, it has reigned as a tribute due to nature throughout the long lapse of the geologic ages from the first beginnings of life upon our planet. What, then, does this inexorable law of death mean? and on what principle does it depend?

It was in mere cobweb toils that those Sadducees who believed 'not in angel, neither in spirit,' endeavoured to entangle our Saviour, when they propounded to him the case of the woman with the seven husbands, and demanded whose wife of the seven she was to be in the Resurrection. But there was a profundity in the reply, which the theologians of nearly two thousand years have, I am disposed to think, failed adequately to comprehend. 'The children of this world marry and are given in marriage,' he said, 'but the children of the Resurrection neither marry nor are given in marriage, neither can they die any more.' Now there seems to be a strictly logical sequence between the two distinct portions of this proposition,—the enunciation that the denizens of the state after death do not marry, and the enunciation that they do not die, which for eighteen centuries there was not science enough in the world adequately to appreciate. The marriage provision was simply a provision tantamount to the original injunction, not of paradise merely, but of every preceding period in which there were organizations of matter possessed by the vital principle: 'Increase and multiply, and replenish the earth.' And all geology presses upon us the conviction, so powerfully enforced by the Liassic deposit at Eathie, that, from the very nature of things, the law of generation and the law of death, wherever space is limited, cannot be dissociated. Each of the multitudinous leaves of the Lias formed in succession an upper surface or platform, on which, for a certain period of time in the world's history, living and sentient creatures pursued the several instincts of their natures, and then ceased to exist. And so immense in many instances was the crowd, that, had the existence of two platforms been restricted to the occupancy of only one platform, they would have lacked footing. A dense crowd of living men may find ample standing-room in an ancient city churchyard, occupying, as they do, a different stratum of space from that occupied by the dead; but were the dead to revive and arise, it would be impossible that the living could find in it the necessary standing-room any longer. They would be jostled from their places far beyond the limits of the enclosing wall. And let us remember, that 'the great globe itself which we inherit' is all one vast burying-ground; nor is it to one stratum that the densely piled remains of its dead are restricted, nor to one hundred, nor to one thousand, nor yet to one hundred thousand strata. Even in this deposit of the Eathie Lias, the successive platforms of the dead may be reckoned up by thousands and tens of thousands; and it would be more possible that a fertile field should have growing upon it at once the harvests of ten thousand succeeding autumns, than that any one of the platforms should have living upon it at once the existences of all the innumerable platforms above and below. The great law *increase and multiply* gave to each platform its countless crowds; and to make room for the continuous operation of this law, the other great law of death came into action, and so the generations of succeeding periods found space to pursue their various instincts on platforms composed in no small part of the perished generations from which they had sprung. Throughout the whole incalculable past of our planet,—throughout all its unmeasured and unmeasurable periods,—the laws of production and decay have gone inseparably together; they were twin stars on the horizon, tinged by the complementary colours, and so inseparably associated, that the appearance of the one always heralded the rise of the other. And, to my mind at least, it does seem demonstrative of the full-orbed and perfect wisdom of the Divine Master of the Theologians, that He, with that quiet simplicity which Pascal so well designates the characteristic style of Godhead, and with a logic too profound to be appreciated at the time, should have coupled together the twin laws of production and decay, as equally inadmissible into that future state in which the life of man is to be no longer

'Summed up in birthdays and in sepulchres.'

'The children of the resurrection neither marry nor are given in marriage, neither can they die any more.'

From the Oolite, with its Liassic base, we pass on to the Triassic system,—a deposit less characteristically developed in England than on the Continent, but of much economic importance, from those vast beds of rock-salt which, in Britain at least, are exclusively restricted to this system; and of considerable geographic importance, from its great lateral extent. In Scotland[1] it occupies rather more than a hundred square miles of surface, chiefly in Dumfriesshire,

[1] There is good reason to believe that the red rocks overlying the coal of Cumberland, the red sandstones of Corncockle Muir, near Dumfries, the Ayrshire red sandstones, and those of the Isle of Arran, are all of the *Permian*, not Triassic, epoch. See *Siluria*, new edition, p. 351.—W.S.S.

along the northern shores of the Solway, and in the line of boundary between the two kingdoms, where it can boast, among its other celebrities, of the famous village of Gretna Green, and the whole of Gretna parish. In England it is chiefly remarkable, in a scenic point of view, for its extreme flatness: its *main feature* is a want of *features*. It was, however, at one time notorious for its ponds and marshes, consequents of the imperfect drainage incident to flat low surfaces when of great extent; and in Scotland, though so much more limited in area, it bears this character still. No fossil organisms have yet been found in this deposit in Scotland: it contains, however, in abundance, traces of the ancient inhabitants, even more curiously imprinted on the stone than if they had left in it remains of their framework; and is interesting as the field in which, from the sedulous study of these, and undeterred by the scepticism of some of our highest authorities, the late Dr. Duncan of Ruthwell laid the first foundations of that curious and instructive department of geologic science since known as *Ichnology*. The strange reptiles of this ancient time, in passing over the tide-uncovered beaches of the district, left their footsteps imprinted in the yielding sand; and in this sand, no longer yielding, but hardened long ages ago into solid rock, the footsteps still remain. And with truly wonderful revelations,—revelations of things the most evanescent in themselves, and of incidents regarding which it might seem extravagant to expect that any record should remain, do we find these strange markings charged. They even tell us how the rains of that remote age descended, and how its winds blew.

Let us see whether we cannot indicate a few of at least the simpler principles of this department of science. The artificial sheet of water situated among the Pentlands, and known as the Compensation Pond, was laid dry, during the warm summer of 1842, to the depth of ten fathoms; and as a lake bottom, ten fathoms from the surface, is not often seen, I visited it, in the hope of acquiring a few facts that might be of use to me among the rocks. What first struck me, in surveying the brown sun-baked bottom from the shore, was the manner in which it had cracked, in the drying, into irregularly polygonal partings, and that the ripple-markings with which it was fretted extended along only a narrow border, where the water had been shallow enough to permit the winds or superficial currents to act on the soft clay beneath. As I descended, I found the surface between the partings indented with numerous well-marked tracks of the feet of men and animals, made while the clay was yet soft, and now fixed in it by the drying process, like the mark of the stamp in an ancient brick. And some of these tracks were charged with little snatches of incident, which they told in a style remarkably intelligible and clear. At one place, for instance, I found the footprints of some four or five sheep. They struck out towards the middle of the hollow, but turned upwards at a certain point, in an abrupt angle, towards the bank they had quitted, and the marks of increased speed became palpable. The prints, instead of being leisurely set down, so as to make impressions as sharp-edged as if they had been carved or modelled in the clay, were elongated by being thrown out backwards, and the strides were considerably longer than those in the downward line. And, bearing direct on the retreating footprints from the opposite bank, and also exhibiting signs of haste, I detected the track of a dog. The details of the incident thus recorded in the hardened mud were complete. The sheep had gone down into the hollow shortly after the retreat of the waters, and while it was yet soft; and the dog, either acting upon his own judgment, or on that of the shepherd, had driven them back. A little further on I found the prints of a shoed foot of small size. They passed onwards across the hollow, the steps getting deeper and deeper as they went, until near the middle, where there were a few irregular steps, shorter, deeper, and more broken than any of the others; and then the marks of the small shoes altogether disappeared, and a small naked foot of corresponding size took their place, and formed a long line to the opposite bank. In this case, as in the other, the details of the incident were clear. Some urchin, in venturing across when the mud was yet soft and deep, after wading nearly half the way shod, had deemed it more prudent to wade the rest of it barefoot, than to bemire his stockings. In each case the incident was recorded in peculiar characters; and to read such characters aright, when inscribed upon the rocks, forms part of the proper work of the ichnologist. His key, so far at least as mere incident is concerned, is the key of circumstantial evidence; and very curious events, as I have said,—events which one would scarce expect to find recorded in the strata of ancient systems,—does it at times serve to unlock.

In some remote and misty age, lost in the deep obscurity of the unreckoned eternity that hath passed, but which we have learned to designate as the Triassic period, a strangely formed reptile, unlike anything which now exists, paced slowly across the ripple-marked sands of a lake or estuary.[1] It more resembled a frog or toad than any animal with which we are now acquainted; but to the batrachian peculiarities it added certain crocodilean features, and in size nearly rivalled one of our small Highland oxen. The prints it made very much resembled those of a human hand; but, as in the frog, the hinder paws were fully thrice the size of the fore ones; and there was a gigantic massiveness in the

---

[1] Reptiles are known to have existed from the period of the Old Red Sandstone, where their tracks have lately been discovered. The reptiles of the Coal are of the Batrachian type; the Permian reptiles are allied to Batrachians and Monitors; while the reptiles of the Trias are Labyrinthodont.—W. S. S.

fingers and thumb, which those of the human hand never possess. Onward the creature went, slowly and deliberately, on some unknown errand, prompted by its instincts; and as the margin of the sea or lake, lately deserted by the water, possessed the necessary plasticity, it retained every impression sharply. The wind was blowing strongly at the time, and the heavens were dark with a gathering shower. On came the rain; the drops were heavy and large; and, beaten aslant by the wind, they penetrated the sand, not perpendicularly, as they would have done had they fallen during a calm, but at a considerable angle. But such was the weight of the reptile, that, though the rain-drops sank deeply into the sand on every side, they made but comparatively faint impressions in its footprints, where the compressive effect of its tread rendered the resisting mass more firm. 'We have here, in a single slab,' says Dr. Buckland, in referring, in his address to the Geological Society for 1840, to these very footprints, and their adjuncts,—'we have here, in a single slab, a combination of proofs as to meteoric, hydrostatic, and locomotive phenomena, which occurred at a time incalculably remote, in the atmosphere, the water, and the movements of animals, from which we infer, with the certainty of cumulative circumstantial evidence, the direction of the wind, the depth and course of the water, and the quarter towards which the animals were passing. The latter is indicated by the direction of the footsteps which form the track; the size and the curvature of the ripple-marks on the sand, now converted into sandstone, show the depth and direction of the current; while the oblique impressions of the rain-drop register the point from which the wind was blowing at or about the time when the animals were passing.'[1] There is another scarce less curi-

[1] The *Labyrinthodon Bucklandi* (Lloyd), formerly believed to be a Triassic reptile, is now ranked as belonging to the *Permian* fauna. See Ramsay, *Quart. Jour. Geol. Soc.* vol. ii. p. 198.—W. S. S.

ous or less minutely recorded incident inscribed on a slab of the same formation, figured and described by Sir Roderick Murchison. It is impressed by the footprints of some be-tailed batrachian, greatly less bulky than the other, that went waddling along much at its leisure, like the sheep in the nursery rhyme, 'trailing its tail behind it.' There is a double track of footprints on the slab,—those of the right and left feet; in the middle between the two, lies the long groove formed by the tail,—a groove continuous, but slightly zig-zagged, to indicate the waddle. The creature, half-way in its course, lay down to rest, having apparently not much to do, and its abdomen formed a slight hollow in the sand beneath. Again rising to its feet, it sprawled a little, and the hinder part of its body, in getting into motion, fretted the portion of the surface that furnished what we may term the fulcrum of the movement, into two wave-like curves. Here, again, are we furnished, from the most remote antiquity, with a piece of narrative of a kind which assuredly we could scarce expect to find enduringly recorded in the rocks. Various reptiles have left curious passages of their history of this kind inscribed on the sandstones of Dumfriesshire; and as Sir William Jardine, the proprietor of some of the quarries, has set himself to the work of illus-tration, the geologist may soon hope to be put in possession of a monograph at once worthy of the subject and of so distinguished a naturalist.[1] The footprints first observed by Dr. Duncan were chiefly those of tortoises; but there also exist in the rock numerous tracks of the huge batra-chians of the period, with traces of a small animal, scarce larger than a rat; and of a nameless, nondescript creature, whose footprints might at the first glance almost be mistaken for those of a horse, but the marks of whose toes have been traced, in some of the impressions, outside the ring of the

---

[1] See Sir William Jardine's work on the *Ichnology of Annandale.*

apparent hoof.[1] And this is all we yet know of these reptilian Triassic inhabitants of Scotland. Robinson Crusoe has gone down to the sea-shore, and seen, much to his astonishment, the print of a savage foot in the sand.

According to an old, but not very old, style of nomenclature, derived from mineralogical character not yet wholly obsolete, the two systems, Triassic and Permian, used to be included under one general head, as the New Red Sandstone, or the Bunter Sandstone of Werner and Jameson. And certainly the mere mineralogist might find it no easy matter to draw a line between them. Up to a certain point in the ascending scale there occurs on the Continent strata of a Red Sandstone known as the Lower Bunter; and immediately over it, a Red Sandstone known as the Upper Bunter.[2] They lie conformably to each other, as if they had been deposited in immediate succession in a still sea: there are no traces of physical convulsion;—the earthquake and the tornado had slept at the time: there was no devastating inundation of molten fire, nor overwhelming wave of translation,—

'It was not in the battle; no tempest gave the shock:'

and yet that undisturbed horizontal line marks where one creation ended and another began. It was held at one time that there was not a single organism, vegetable or animal, common to the two great divisions to which these sandstone beds belong; but there now seems to rest some doubt on the point. In an insulated district of France, plants of the Coal Measures have been found in a deposit containing Belemnites; and it is held that the Belemnite belongs exclusively to the great Secondary division. But

[1] *Chelichnus Titan* and *Gigas Jardine.*—W. S. S.

[2] The Bunter sandstein and Bunter schiefer; of which the Bunter sandstein is now ranked as lowest, *Trias*, and the Bunter schiefer as upper, *Permian.*—W. S. S.

the specific standing of these Belemnites still remains to be determined: it is possible they may not be Secondary forms; and it has been suggested by M. Michelin, a distinguished French geologist, that generically the Belemnite may not be of the premised importance in reference to the age of these Tarentaise beds. 'He is inclined,' we find him saying, 'to consider it an instance of the occurrence of the Belemnite form in the Carboniferous period, rather than of the continuance of the same *species* of plants through several sucessive epochs.'

But, leaving it to the future researches of geologists to determine whether there be any, and, if so, how many, organisms common to the Secondary and Palæozoic divisions, a very slight acquaintance with fossils is sufficient to show that between the types of organic nature in these two great divisions there exist differences and distinctions of the broadest and most palpable kind. In passing upwards from the Triassic to the Permian, we seem to pass, not merely from one dynasty to another, but, if I may dare employ such a term, from one dispensation to another. So broad are the differences, that they affect whole classes of the animal kingdom. In the class of fishes, for instance, an entire change takes place in the form of the tail. There are a few ichthyic families in the present day, such as the sharks and sturgeons, that have unequally-lobed tails, from the circumstance that a prolongation of the vertebral column runs into the upper lobe; whereas in perhaps nineteen-twentieths of all the existing families, the vertebral column stops short, as in the osseous fishes common at our tables, a little over the lobes, to form for them a medial basis, and the lobes themselves are equal. And this equal-lobed, or, as it is termed, *homocercal* condition, is the prevailing one, not only in the present time, but in all the Tertiary and in all the Secondary ages, up till the close of the Triassic system. And then, sudden as the shifting of a scene, or as one of the abrupt transitions of a dream, we find, immediately on entering the great Palæozoic division, an entire change. The unequal-lobed or *heterocercal* tail becomes not only the prevailing, but the only form, save in a few exceptional cases, as in that of the *Coccosteus* of the Old Red Sandstone, where there were no lobes at all, or as in that of its contemporary the *Diplopterus*, where the lobes strike out laterally from a prolongation of the column. In short, the equally-lobed tail ceases with the Trias, to re-appear no more, and the unequally-lobed tail takes its place. Similar changes manifest themselves in other divisions and classes of the animal kingdom. Waiving for the present the question raised by the French geologist, M. Michelin, in Britain at least the Belemnite, so abundant in the Secondary formations, and so characteristic of them, has no place among the formations of the Palæozoic period. Save, too, in a few rare and somewhat equivocal species, the equally characteristic Ammonite disappears.[1] We take leave also of the scarce less characteristic Gryphites, of the Trigonia, Plagiostoma, and Perna, with several other well-marked types of shell; but we find their places amply occupied by types exclusively Palæozoic. The Orthoceratites, straight, conical, chambered shells, anticipated, we see, the place of the Belemnites ; the Goniatites, that of the Ammonites proper; the Bellerophon and the Euomphalus, unseen in any other period, fall into the general group, and add to the peculiarity of its aspect; with a whole array of unwonted forms among the brachiopoda, such as Spirifers, Producta, Atrypa, and Pentamerus, etc. etc. But it was perhaps in the vegetable world that the Palæozoic ages most remarkably differed from those of the subsequent periods of the geologist, whether Secondary, Tertiary, or Recent. We read in the older poets of enchanted forests; but the true enchanted forests, stranger, in their green luxuriance, than poet ever yet fancied, and where the botanist,

[1] These views require much modification. See Sir Charles Lyell's *Supplements*, 1857.—W. S. S.

surrounded by irreduceable shapes that would take no place in his systems, might well deem himself in a wild dream, were the forests of the Coal Measures.

The Coal Measures of Scotland occupy about two thousand square miles of surface, and, though much overflown by igneous rock, and occasionally broken through by patches of Old Red Sandstone, run diagonally across the country, from sea to sea, in a tolerably well-defined belt, nearly parallel to the line of the southern flank of the Grampians. Throughout their entire extent they owe their scenic peculiarities to the trap ; but where least disturbed, as in the Dalkeith coalfield, they are of an inconspicuous, low-featured character, and chiefly remarkable for their rich fields, as to the east of Edinburgh, between the Arthur Seat group of hills and the Garleton hills near Haddington ; or for their romantic dells and soft pastoral valleys, such as Dryden Dell, or the valley of Lasswade, or to enumerate two other well-known representative localities in one stanza, borrowed from Macneil,

> 'Roslin's gowany braes sae bonny,
> Crags and water, woods and glen ;
> Roslin's banks, unpeer'd by ony,
> Save the Muse's, Hawthornden.'

The coal-fields owe some of their more characteristic features, especially in the sister kingdom, to man. The tall chimneys, ever belching out smoke ; the thickly-sown engine-houses, with the ever-recurring clank of the engines, and the slow-measured motion of their outstretched arms seen far against the sky ; the involved fretwork of railways, connected with some main arterial branch, along which the traveller ever and anon marks the frequent train sweeping by, laden with coals for the distant city ; the long flat lines of low cottages, the homes of the poor colliers ; and here and there, where the ironstone bands occur, a group of smelting furnaces ;—all serve to mark the Coal Measures, and to distinguish them from every other system. And such—striking off the peculiarities of the trap, which has no necessary connexion with the Carboniferous system, but is common, in some one part of the world or another, to all the systems—are some of the features, natural and superinduced, of this most important, in an economic point of view, of all the geologic formations. They are, as I have said, of no very prominent character. The poet Delta describes, in a fine stanza, the scenery around and to the east of Edinburgh. But though the area which the landscape includes contains one of our most considerable coal-basins,—a basin many square miles in extent, —it does not furnish him with a single descriptive reference. Almost all those bolder and more characteristic features of the scene which his pencil exquisitely touches and relieves, it owes to the igneous rocks :

> 'Traced in a map the landscape lies,
>   In cultured beauty stretching wide ;
> There Pentland's green acclivities ;
>   There ocean with its azure tide ;
> There Arthur Seat, and, gleaming through
>   Thy southern wing, Dunedin blue ;
> While in the orient, Lammer's daughters,
>   A distant giant range, are seen ;
> North Berwick Law with cone of green,
>   And Bass amid the waters.'

The ancient scenery of the Coal Measures would be greatly more difficult to trace. As we recede among the extinct creations farther and farther from the present time, the forms become more strange, and less reduceable to those compartments to which we assign known classes and existing types. Our more solid principles of classification desert us, and we are content to substitute instead, remote analogies and distant resemblances. We say of one family of plants that they somewhat resemble club-mosses, shot up in bulk and height into forest trees ; and of another family, that they would be not very unlike the horsetails of our morasses, did horsetails rival in size larches of some twenty or thirty years' growth. In referring to yet other families, we can avail ourselves— so *outré* are their forms—of no resemblance at all: we can simply figure and describe, and draw our illustrative comparisons, if we employ such, rather from the departments of art than of nature. It is possible that, were some of our higher botanists—our Balfours, Browns, and Grevilles—permitted to range for a day over the broad plains of Jupiter, or amid the bright sunshiny vales of Mercury or Venus, even *they* might be but able to tell us, on their return, of gorgeous floras, that defied all their old rules of classification, and which could be illustrated from that of our own planet only by distant resemblances and remote analogies. And assuredly such would be the case, could they, through the exercise of some clairvoyant faculty, be enabled to journey for millions and millions of years into the remote past, and to spend a few enchanted hours amid the dense and sombre thickets of a Carboniferous forest. Shall I venture on communicating to this audience a snatch of personal history, illustrative of the mode in which I myself arrived, many years ago, at my earliest formed conceptions regarding the old flora of the Coal Measures?

The first perusal of *Gulliver's Travels* forms an era in the life of a boy, if the work come in his way at the right time; and I was fortunate enough to secure my first reading of it at the mature age of eight years. For weeks, months, years after, my imagination was filled with the little men and little women, and with at least one scene laid in the country of the very tall men,—the scene in which Gulliver, after wandering amid grass that rose twenty feet over his head, lost himself in a vast thicket of barley forty feet high. I became the owner, in fancy, of a colony of little men: I had little men for inhabiting the little houses which I built, for tilling my little apron-breadth of a garden, and for sailing my little ship; and, coupling with the men of Lilliput the scene in Brobdingnag, I often set myself to imagine, when playing truant all alone on the solitary slopes or amid the rocky dells of Drieminorie, how the little creatures, who were sure always to accompany me on these occasions, would be impressed by the surrounding vignette-like scenes and mere picturesque productions, exaggerated on hill and in hollow, by their own minuteness, into great size. I have imagined them threading their way through dark and lofty forests of bracken fifty feet high, or admiring on the hill-side some enormous club-moss, that stretched out its green hairy arms over the soil for whole roods, or arrested at the edge of some dangerous and dreary morass by hedges of gigantic horsetail, that bore atop their many-windowed, club-like cones, twenty feet over the dank surface, and that shot forth at every joint their green verticillate leaves in rings huge as coach-wheels. And while I thus thought, or rather dreamed, for my Lilliputian companions, I became for the time a Lilliputian myself,—saw the minute in nature as if through a magnifying-glass,—roamed in fancy under ferns that had shot up into trees,—and saw the dark cones of the Equisetaceæ stand up over their spiky branches some six yards or so above head. But these day-visions belonged to an early period: dreams of at least a severer, if not more solid cast, dispossessed the little men and women of the place they had occupied; and I had learned to think of the wondrous tale of Swift as one of the most powerful but least genial of all the satires which the errors and perversions of poor human nature have ever provoked, when in the year 1824 I formed my first practical acquaintance with the flora of the Coal Measures. I was engaged as a stone-cutter, a few miles from Edinburgh, in making some additions in the old English style to an ancient mansion-house; and the stone in which I wrought,—a curiously variegated sandstone derived from a quarry since shut up,—was, I soon found, exceedingly rich in organic casts and impressions.

They were exclusively vegetable. Often have I detected in the rude block placed before me, to be fashioned into some moulded transom or carved mullion, fragments of a sculpture which I might in vain attempt to rival,—the forked stems of Lepidodendra, fretted into scales that, save for their greater delicacy and beauty, might have reminded the antiquary of the sculptured corslet of scale-armour on the effigies of some ancient knight; the straight-stemmed Calamite, fluted from joint to joint, like the shaft of some miniature column of the Grecian Doric; the Sigillaria, also a fluted column, but of a more meretricious school than that of Greece, for it was richly carved between the flutings; the Stigmaria, fretted over, with its eyelet-holes curiously connected by delicately-waved lines; and occasionally the elaborately ornate Ulodendron, with its rows of circular scars, that seemed to have been subjected to the lathe of an ornamental turner, and its general surface fretted over with what seemed to be nicely sculptured leaves, such as we sometimes see on a Corinthian torus. It was not easy, more than a quarter of a century ago, when Sir Roderick Murchison was still an officer of dragoons, Sir Charles Lyell prosecuting the study of English law, and Dr. Buckland still engaged with his theory of the Flood, which he had given to the world only the previous year,—it was not easy, I say, for a working man to have such questions solved as these fossils of the Coal Measures served to raise. But they *were* at length in some measure solved. I was taught to look to those forms of the existing flora of our country that most resembled the forms of its flora during the Carboniferous period. And, strange to tell, I found I had just to fall back on my old juvenile imaginings, and to form my first approximate conceptions of the forests of the Coal Measures by learning to look at our ferns, club-mosses, and equisetaceæ, with the eye of some wondering traveller of Lilliput lost amid their entanglements, like Gulliver among those of the fields of Brobdingnag. When sauntering, after the work of the day was over, along the edge of some wood-embosomed streamlet, where the horsetail rose thick and rank in the danker hollows, and the fern shot out its fronds from the drier banks, I had to sink in fancy, as of old, into a mannikin of a few inches, and to see intertropical jungles in the tangled grasses and thickly interlaced equisetaceæ, and tall trees in the herbaceous plants and the shrubs.

But many a wanting feature had to be supplied, and many an existing one altered. Amid forests of arboraceous ferns, tall as our second-class trees, there stood up gigantic club-mosses thicker than the body of a man, and from sixty to eighty feet in height; more than a hundred and fifty species of smaller ferns, and about one-third that number of smaller species of club-mosses, clothed the opener country; and along the frequent marshes and lakes that covered vast tracts of its flat surface, or the sluggish rivers that winded through it, there flourished huge thickets of equisetaceæ, of from twelve to fourteen different species, tall, some of them, as the masts of pinnaces, and thick and impenetrable as the fairy hedge that surrounded the palace of the Sleeping Beauty. But among these forms of the vegetable world, that, at least through the blue steaming vapour of so dank a land, seem but the more familiar forms of our lochans and hill-sides many times magnified, there arise strange floral shapes, among which we can recognise no existing type. The Ulodendron, bearing along its carved trunk, on two of its sides, rectilinear strips of cones, like rows of buttons on the dress of a boy, and the ornately tatooed Sigillaria, lined longitudinally, and with its thickly-planted vertical rows of leaves bristling from its stem and larger boughs, resemble no vegetable productions which the earth now yields. The landscape, too, has its intertropical forms, —what seem gigantic Cacti, with thickets of canes, and a few species of palms. And, where here and there a flat hillock rises a few yards over the general level, we see groups of noble Araucarians raising their green tops a hundred and fifty feet over the plain. And yet, rich as the flora of the period may seem in individuals, and though it cumbers the soil with a luxuriance witnessed in our own times only among the minuter forms, it is, in all save size and bulk, a poor and low flora after all. The Pines and Araucarians form its only forest-trees. We fail to meet on its plains a single dicotyledonous plant on which a herbivorous mammal could browse. Its Lycopodiaceæ are covered over with catkin-like cones; thére are cones on its Ulodendra, cones on its Equisetaceæ, cones on its Araucarians, cones on its Pines; but not a single fruit have we yet found good for the use of man. Nor, after the first impression of novelty has passed away, is there much even to gratify the sight. The marvel of ornately-carved trunks and well-balanced fronds soon palls on the sense; and the prevalence of those spiky rectilinear forms in the scene which Wordsworth could regard as such deformities in landscape, and which James Grahame so deprecates in his *Georgics*, 'lies like a load on the weary eye.' Nature labours in the productions of huge immaturities; neither man the monarch, nor his higher subjects the mammals, have yet appeared; and it is all too palpable that that garden has not yet been planted, out of the ground of which there shall grow 'every tree that is pleasant to the sight and good for food.'

Some of the gigantic forms of these primeval forests we can only vaguely and imperfectly illustrate by the dwarf productions of our present moors and morasses; and some of them we fail to connect, by the links of general resemblance, with aught in the vegetable kingdom that now lives. Regarded as a whole, the flora of the Carboniferous age seems as remote in its analogues from that which now exists, as remote in the period during which it flourished.

There are, however, at least two families of plants which bear, not a loose and general, but a minute and thorough resemblance, to families which also existed during the great Secondary and Tertiary periods, and which still continue to occupy a large space among the recent vegetable forms. And these are the Fern and the Pine families. All the species have become extinct over and over again; but the families, and many of the genera, are ever reproduced; and, so far as we know, this earth never possessed a terrestrial flora that had not its ferns and its pines. In all the other divisions and classes of the organic world there are also favourite families, such as the Tortoises among reptiles, the Cestracions among fishes, the Nautilus among Cephalopodes, and the Terebratula among Brachipods. There are few geologic formations in which either the remains or the footprints of tortoises have not been detected; there seems never to have been an ocean that had not its cestracion; the nautilus lived in every age from the times of the Lower Silurian deposits down to the present day; and, after disinterring specimens of fossil terebratula from our Grauwackes, our Mountain Limestones, our Oolites, and our Chalk Flints, we may cast the drag in the deeper lochs of the Western Highlands, and bring up the living animals, fast anchored by their fleshy cables to stones and shells. We can scarce glance over a group of fossils of the two earlier divisions, the Secondary and the Palæozoic, which we do not find divisible into two classes of types,—the types which still remain, and the types which have disappeared. But why the one set of forms should have been so repeatedly called into being, while the other set was suffered to become obsolete, we cannot so much as surmise. In visiting some old family library that has received no accession to its catalogue for perhaps more than a century, one is interested in marking its more vivacious classes of works,—its Shakespeares, Robinson Crusoes, and Pilgrim's

Progresses,—in their first, or at least earlier editions, ranged side by side with obsolete, long-forgotten volumes, their contemporaries, with whose unfamiliar titles we cannot connect a single association. And exactly such is the class of facts with which the geologist is called on to deal. He finds an immense multiplication of editions in the case of some particular type of fish, plant, or shell; and in the case of other types, no after instances of republication, or republication in merely a few restricted instances, and during a limited term. But while it is always easy to say why, in the race of editions, the one class of writings should have been arrested at the starting-post, and the other class should go down to be contemporary with every after production of authorship until the cultivation of letters shall have ceased, the geologist finds himself wholly unable to lay hold of any critical canon through which to determine why, in the organic world, one class of types should be so often republished, and another so peremptorily suppressed. This far, however, we may venture to infer, from finding the two classes under such a marked diversity of dispensation, that creation must have been a result, not of the operation of mere law, which would have dealt after the same fashion with both, but a consequence of the exercise of an elective will; and that as amid immense variety of effort and fertility of invention there are yet certain features of style, and a certain recurrence of words and phrases, that enable us to identify a great author, and to recognise a unity in his works that bespeaks the unity of the producing mind, so ought these connecting links and common features of widely-separated, and in the main dissimilar creations, to teach us the salutary lesson that the Author of all is One, and that, in the exercise of his unrestricted sovereignty and of his infinite wisdom, He chooses and rejects according to his own good pleasure.

From the plants of the swamps and forests of the Coal

Measures we pass on to its fauna, terrestrial and aquatic;—a fauna which, although less picturesque than its wondrous flora, filled with all manner of strange shapes, seems to have borne a corresponding character in uniting great numeric development to a development comparatively limited in classes and orders; and with respect also to the extreme antiqueness of many of its types. The prevailing forms of both flora and fauna belong equally to a fashion that has perished and passed away.

It was held, up till a very recent period, that there had existed no reptiles during the Carboniferous ages. Man has been longer and more perseveringly engaged among the Coal Measures than in any of the other formations; and, long ere geology existed as a science, what used to be termed its figured stones,—plants, shells, and fishes,—were, in consequence, well known to *collectors*,—a class of people sent into the world to labour instinctively as pioneers in the physical sciences, without knowing why. I have seen prints of some of these figured stones of two centuries' standing, and have succeeded in recognising as old acquaintance the Spirifers and Ferns which had sat for their pictures to artists who knew nothing of either. During the last sixty years there have been many collections made of the Carboniferous fossils, and many coal-fields intelligently examined, but not a trace of the reptile detected. It was not until Sir Charles Lyell's second visit to the United States, five years ago, or rather not until the publication of his second series of travels, three years after, that it was known to European geologists that the coal-fields of Pennsylvania, in the United States, had, like the Trias[1] of the south of Scotland and of the sister kingdom, their Cheirotherium, of, however, not only, as might be anticipated, a different species, but of even a different genus, from that of the newer formation, though not less decidedly

[1] Permians.—W. S. S.

reptilian in its character. And about the same time the remains of a reptile since known as the Archegosaurus were found in a coal-field in Rhenish Bavaria. The Archegosaurus seems to have been a strange-looking creature,—half-saurian, half-batrachian, of comparatively small size, with two staring eyes set close together in the middle of a flat triangular skull, and furnished with limbs terminating in distinct toes, but so slim and weak, 'that they could have served,' says Von Meyer, 'only for swimming or creeping.'[1] It is stated in the *Lake Superior* of Agassiz, that in a shallow expanse of the river into which the lake falls, skirted by flat forest-covered banks, and in which a long series of dreary mud-flats are covered by from a few inches to a few feet of water, there occurs a large gill-furnished salamander (Menobranchus), which the Indians call the 'walking fish,' and which even to them is a great curiosity. It swims wherever there is sufficient depth of water, and creeps over the mud-flats where there is not; and, compared with the swift and powerful Lepidosteus, a reptile-fish of the same stream, it is a stupid, sluggish, inert creature, safe only in its uselessness and the repulsiveness of its appearance. And, judging from the feebleness of its limbs, and the shortness of its ribs, which resemble, says Professor Owen, those of the half-lunged, half-gilled Proteus, such seems to have been the character of the Archegosaurus. Its contemporary, the American Cheirotherium, as shown by its well-defined footprints, must have been a stronger-limbed and larger reptile,—a batrachian heightened by a dash of the crocodile; and, though probably often a dweller in the water, the only vestiges of it which remain show that it must have occasionally stepped out of its river or lake, to take an airing on the banks. Such is nearly the sum-total of our knowledge regarding the reptiles of the Carbonifer-

[1] The Archegosauri are related to the Batrachians and Sauroid fishes, according to Owen—*Siluria*, new edition, p. 363.—W. S. S.

ous period.[1] Like mammals in the preceding Secondary ages, they formed so inconspicuous a feature of the fauna of the time, that until very recently it escaped notice, and so was not recognised as a feature at all. So far as we yet know, the great Secondary division, in which reptiles, both in size and number, received their fullest development, had but few genera of mammals,—a small pouched animal, and small insectivorous ones: so far as we yet know, the great Palæozoic division, in which fishes, both in size and number, received their fullest development, had but its two genera of reptiles, both allied, apparently, to the humble batrachian order. The reigning dynasty of the one period, though the mammal was present, was not that of the mammal, but of the reptile: the reigning dynasty of the other period, though the reptile was present, was not that of the reptile, but of the fish.

The fishes of the Coal Measures, in especial the reptile fishes, were in truth very high types of their class. I have already incidentally said, that with the humble Menobranchus or salamander of the great North American lakes and their tributaries, there is a true reptile fish associated;—an order of creatures of which, so far as is yet known, there exists in the present creation only a single genus. It would almost seem as if the Lepidosteus had been spared, amid the wreck of genera and species, to serve us as a key by which to unlock the marvels of the ichthyology of those remote periods of geologic history appropriated to the dynasty of the fish. This wonderful creature is covered by scales, not of a horny substance, like those of the fish com-

---

[1] Lord Enniskillen possesses a fossil reptile allied to the Cheirotherium from the Yorkshire coal-fields, the Parabatrachus Colei (Owen). A Labyrinthodont reptile, *Baphates planiceps* (Owen), has been found in the Nova Scotia coal-fields. Also footmarks of sauroid reptiles have been discovered in Scotland by Mr. Hugh Miller, and in the Forest of Dean by Mr. C. Bromby.—W. S. S.

mon at our tables, but of solid bone, enamelled, like the human teeth, on their outer surfaces. Its own teeth are planted in double rows of unequal size, the larger being of a reptilian, the smaller of an ichthyic character; and the front teeth of the lower jaw are received, as in the alligator, into sheath-like cavities in the upper jaw,—another reptilian trait. Its vertebral column, wholly unlike that of other fishes, each of whose vertebræ consists of a double cup, is formed of vertebræ one end of which consists of a cup and another of a ball,—a characteristic of the snake: it possesses true gills, like all other fishes; but then it also possesses a peculiar form of cellular air-bladder, opening into the throat by a glottis, which, according to Agassiz, our highest authority, performs respiratory functions. The Lepidosteus, says Sir Charles Lyell, in describing, in his second series of travels in the United States, an individual which he had seen in sailing across Lake Solitary, leap like a trout or salmon over the surface, in pursuit of its prey,— 'the Lepidosteus, whose hard shining scales are so strong and difficult to pierce that it can scarcely be shot, can live longer out of the water than any other fish of the United States, having a large cellular swimming-bladder, which is said almost to serve the purpose of a real lung.' Further, we find Agassiz stating, in his *Lake Superior*, that the Lepidosteus is one of the swiftest of fishes, darting like an arrow through the waters, and overcoming with facility even the rapids of the Niagara. He adds further, that when at the latter place, there was a living specimen caught for him,— the first living specimen he had ever seen; and that 'to his great delight, as well as to his utter astonishment, he saw this fish moving its head upon its neck freely, right and left, and upwards, as a saurian, and as no other fish in creation does.' The true native Yankee has a mode wholly his own, and somewhat redolent of the revolver and the bowie-knife, of describing the peculiar immunities and high standing of the Lepidosteus, or, as he familiarly terms it, the gar-pike. 'The gar-pike is,' he says, 'a happy fellow, and beats all fish-creation: he can hurt everything, and nothing can hurt him.' And such is the living type of what was the prevailing and dominant family of the fauna of the Coal Measures.

The great size and marvellous abundance of those reptile fishes of the Carboniferous period may well excite wonder. One ironstone band in the neighbourhood of Gilmerton has furnished by scores, during the last few years, jaws of the *Rhizodus Hibberti* and its congeners, of a mould so gigantic, that the reptile teeth which they contain are many times more bulky than the teeth of the largest crocodiles. Teeth and scales of the same genus are also abundant among the limestones of Burdiehouse;—some of the teeth much worn, as if they had belonged to very old individuals; and some of the scales, which were as largely imbricated as those of the haddock or salmon, full five inches in diameter. The broken remains of a Burdiehouse specimen now in the museum of the Royal Society of Edinburgh are supposed by Agassiz to have formed part of one of the largest of true fishes,—a fish which might be appropriately described in the sublime language applied in Job to Leviathan. If the gar-pike, a fish from three to four feet in length, can make itself so formidable, from its great strength and activity, and the excellence of its armour, that even the cattle and horses that come to drink at the water's side are scarce safe from its attacks, what must have been the character of a fish of the same reptilian order from thirty to forty feet in length, furnished with teeth thrice larger than those of the hugest alligator, and ten times larger than those of the bulkiest Lepidosteus, and that was covered from snout to tail with an impenetrable mail of enamelled bone? 'Canst thou play with Leviathan as with a bird? Canst thou fill his skin with barbed irons, or his head with fish-spears? Who can open the doors of his face? His teeth are terrible round about; his scales are his pride, shut up together as a close seal. In his neck remaineth strength; his heart is as firm as a stone, yea, as hard as a piece of the nether millstone. The sword of him that layeth at him cannot hold; the spear, the dart, nor the habergeon. He esteemeth iron as straw, and brass as rotten wood.'

In the same waters as the formidable and gigantic Holoptychian genus there lived a smaller but still very formidable reptile fish, now known as the *Megalichthys*,—a fish whose body was covered with enamelled quadrangular scales, and its head with enamelled plates, both of so exquisite a polish, that they may still be occasionally seen in the shale of a coal-pit, catching the rays of the sun, and reflecting them across the landscape, as is often done by bits of highly glazed earthenware or glass. It was accompanied by another and still smaller, but very handsome, and scarce less highly enamelled, genus of the sauroid class,—the *Diplopterus*. And if, after the lapse of so many ages, their armour still retains a polish so high, we may be well assured that brightly must it have glittered to the sun when the creatures leaped of old into the air, like the Lepidosteus of Lake Solitary, after some vagrant ephemera or wandering dragon-fly; and brightly must the reflected light have flashed into the dark recesses of the old overhanging forest that rose thick and tangled over the lake or river side. The other ichthyic contemporaries of these fishes were very various in size and aspect. About half their number belonged to the same ganoidal or bone-covered order as the Holoptychius and Megalichthys, and the other half to that placoidal order represented in our existing seas by the sharks and rays. The lakes, rivers, and estuaries abounded, perhaps exclusively, in ganoids, such as the *Palæoniscus*, a small, handsome, well-proportioned genus, containing several species, —the *Eurynotus*, a rather longer and deeper genus, formed somewhat in the proportions of the modern bream,—and the *Acanthodes*, an elongated, spined, small-scaled genus, formed in the proportions of the ling or conger eel. On the other hand, the seas of the period, abundant also in ganoids, were tenanted by numerous and obsolete families of sharks, amply furnished both with razor-like teeth in their jaws for cutting, and millstone-like teeth on their palates for crushing,—furnished, some of them, with barbed stings, like the sting-rays,—and whose dorsal fins were armed with elaborately carved spines. The only representative of any of these genera of marine placoids which still exists is the Cestracion or Port-Jackson shark, a placoid of the southern hemisphere.

We know that over the rivers and lakes inhabited by the ganoidal fishes of this period there fluttered several species of insects mounted on gauze wings, like the Ephemeridæ of the present day. At least one of their number must have been of considerable size;—a single wing preserved in ironstone, though not quite complete, is longer than the anterior wing of one of our largest dragon-flies, and about twice as broad; and, as its longitudinal nervures are crossed at nearly right angles by transverse ones, it must have resembled, when attached to the living animal, a piece of delicate network. In the woods, and among the decaying trunks, there harboured at the same time several species of snouted beetles, somewhat akin to the diamond beetles of the tropics; and with these, large many-eyed scorpions. The marshes abounded in minute crustaceans, of, however, a low order, that bore their gills attached to their feet, and breathed the more freely the more merrily they danced; and the seas contained the last of the trilobites. I have already referred incidentally to the shells. The fresh waters contained various forms of Unio, somewhat similar to the pearl mussels of our rivers; the profounder depths of the sea had their brachipods,—Spirifers and Producta; while molluscs of a higher order,—Orthoceratites, some of them of gigantic size, Nautilus, and Goniatite, swam above. Corals of strange shapes were abundant : there were several species of Tubilipora, which more resembled the organ-pipe coral than aught else that still exists ; with great numbers of a horn-shaped coral, Turbinolia, with its point turned downwards, like that of a Cornucopia, and with an animal somewhat akin to the sea-anemone, expanded, flower-like, from its upper end. With these, too, there were grouped delicately branched corals, mottled with circular cells; and minutely elegant Fenestrella, that seemed reduced editions of the sea-fan. An antiquely-formed sea-urchin, whose spines were themselves roughened with minute spines, as the more delicate branches of a sweet brier are roughened with thorns, crept slowly among these zoophytes by its many cable-like tentacula ; while forests of Crinoidea waved in the tide, and sent abroad their many arms from the ledges over-head. These forests of Crinoidea or stone-lilies formed one of the leading characteristics of the sea-bottoms of the period. We may conceive of them as thickets of flexible-jointed stems rooted to the rocks, and with a variously-formed star-fish fixed on the top of each stem. Some of the stems were branched, some simple ; in some the petals or rays were richly palmated ; in others, plain and star-like ; in some, what might be deemed the calyx of the flower, but which was in reality the stomach of the animal, was round and polished ; in others, ornately carved into regular geometrical figures. But however various in their appearances, they were all sedentary star-fishes, that, poised on their tall, cane-like stems, sent abroad their arms into the waters of the old Carboniferous ocean, in quest of food. The minute joints of the stem, perforated in the middle by a circular passage, and fretted by thick-set rays radiating from the centre, seem to have attracted notice in an early age, and are known in legendary lore as the beads of St. Cuthbert. Dr. Mantell states that he has found quantities of these perforated ossicula, which had been worn as ornaments, in tumuli of the ancient Britons. And you will remember that in *Marmion*, the nuns of St. Hilda, who lived in a Liassic country rich in Ammonites, had their stories regarding the snakes which their sainted patroness had changed into stone; and that they were curious to know, in turn, from the nuns of Lindisfarne, who lived in a Carboniferous district, rich in encrinites, the true story of the beads of St. Cuthbert:

> ' But fain St. Hilda's nuns would learn,
> If on a rock by Lindisfarne
> St. Cuthbert sits, and toils to frame
> The sea-born beads that bear his name.
> Such tales had Whitby's fishers told,
> And said they might his shape behold,
> And hear his anvil sound.
> A deadened clang, a huge dim form,
> Seen but and heard, when gathering storm
> And night were closing round.'

Certainly, if he fabricated all the beads, he must have been one of the busiest saints in the Calendar. So amazingly abundant were the lily encrinites of the Carboniferous period, that there are rocks in the neighbourhood of Edinburgh, of considerable thickness and great lateral extent, composed almost exclusively of their remains.

The depth of the Carboniferous system has been well described as enormous. Including the Mountain Limestone, which is a marine deposit of the same period, and which must be regarded as forming a member of the Coal Measures, there are districts of England in which, as estimated by Mantell, it has attained to the vast thickness of ten thousand feet. In our own immediate neighbourhood it does not, as estimated by a high authority, Mr. Charles M'Laren, quite equal half that depth. Our Carboniferous system, including the Roslin and Calciferous sandstones, he describes, in his *Geology of Fife and the Lothians*, as about four thousand five hundred feet in thickness,—a thickness, however, which more than equals the height of Ben Nevis over the level of the sea. That coal-basin which extends along the flat richly cultivated plain which stretches from the south-eastern flanks of Arthur Seat to the Garleton Hills in Haddingtonshire, considerably exceeds three thousand feet in depth; and, could it be cleared out to the bottom of the Calciferous sandstones, and divested of the hundred and seventy beds of which it consists, as we have seen the deep hollow of the Compensation Pond divested of its water, it would form by far the profoundest valley in Scotland. Of the beds by which it is occupied, it is estimated that about thirty are coal, varying from several feet to but a few inches in thickness; and we now know, that though some of these coal-seams were formed of drifted plants and trees deposited in the sandy bottom of some great lake or inland sea, by much the greater number are underlaced by bands of an altered vegetable soil, thickly traversed by roots; and that, as in the case of many of our larger mosses, the plants which entered into their composition must have grown and decayed on the spot. And of course, when the plants were growing, the stratum in which they occur, though subsequently buried beneath plummet sound, or at least thousands of feet, must have formed a portion of the surface of the country either altogether sub-aërial, or, if existing as a swamp, overlaid by but a few inches of water. We have evidence of nearly the same kind in the ripple-markings which are so abundant throughout all the shales and sandstones of the Coal Measures from top to bottom, and which are never formed save where the water is shallow. Stratum after stratum in the whole ten thousand feet included in the system, where it is most largely developed, must have formed in succession the surface either of the dry land or of shallow lakes or seas; one bed must have sunk ere the bed immediately over it could have been deposited; and thus, throughout an extended series of ages, a process must have been taking place on the face of the globe somewhat analogous to that which takes place during a severe frost in those deeper lakes of the country that never freeze, and in which the surface stratum, in consequence of becoming heavier as it becomes colder than the nether strata, is for ever sinking, and thus making way for other strata, that cease to be the surface strata in turn. This sinking process, though persistent in the main, must have been of an intermittent and irregular kind. In some instances, forests seem to have grown on vast platforms, that retained their level unchanged for centuries, nay, thousands of years together : in other cases the submergence seems to have been sudden, and to such a depth, that the sea rushed in and occupied wide areas where the land had previously been, and this to so considerable a depth, and for so extended a period, that the ridges of coral which formed, and the forests of Encrinites which grew, in these suddenly hollowed seas, composed thick beds of marine limestone, which we now find intercalated with coal-seams and lacustrine silts and shales. There seem, too, to have been occasional upward movements on a small scale. The same area which had been occupied first by a forest, and then by a lake or sea, came to be occupied by a forest again ; and, though of course mere deposition might have silted up the lake or sea to the level of the water, it is not easy to conceive how, without positive upheaval for at least a few feet, such surfaces at the water-level should have become sufficiently consolidated for the production of gigantic Araucarians and Pines. But the sinking condition was the general one ; platform after platform disappeared, as century after century rolled away, impressing upon them their character as they passed ; and so the Coal Measures, where deepest and most extensive, consist, from bottom to top, of these buried platforms, ranged like the sheets of a work in the course of printing, that, after being stamped by the pressman, are then placed horizontally over one another in a pile. Another remarkable circumstance, which seems a direct result of the same physical conditions of our planet as those ever-recurring subsidences, is the vast horizontal extent and persistency of these platforms. The Appalachian Coal formation in the United States has been traced by Professor Henry Rogers over an area considerably more extensive than that of all Great Britain; and yet there are some of its beds that seem continuous throughout. The great Pittsburg coal-seam of this field,—a seam wonderfully uniform in its thickness, of from eight to twelve feet,—must have once covered a surface of ninety thousand square miles. And this characteristic of persistency, united to great extent, in the various platforms of the Coal Measures, and of ever-recurring subsidence and depression, which accumulated one surface platform over another for hundreds and thousands of feet, belong, I am compelled to hold, to a condition of things no longer witnessed on the face of the globe. The earth has still its morasses, its deltas, its dismal swamps; it has still, too, its sudden subsidences of surface, by which tracts of forest have been laid under water; but morasses and deltas cover only very limited tracts, and sudden subsidences are at once very exceptional and merely local occurrences. Subsidence during the Carboniferous ages, though interrupted by occasional periods of rest, and occasional paroxysms of upheaval, seems, on the contrary, to have been one of the fixed and calculable processes of nature; and, from apparently the same cause, persistent swamps, and accumulations of vegetable matter, that equalled continents in their extent, formed one of the common and ordinary features of the time.

My subject is one on which great diversity of opinion may and does prevail. But while entertaining a thorough respect for the judgment and the high scientific acquirements of geologists who hold that the earth existed at this early period in the same physical conditions as it does now, I must persist in believing that these conditions were in one important respect essentially different; I must persist in believing that our planet was greatly more plastic and yielding than in these later times; and that the molten abyss from which all the Plutonic rocks were derived,—that abyss to whose existence the earthquakes of the historic period and the recent volcanoes so significantly testify,—was enveloped by a crust comparatively thin. Like the thin ice of the earlier winter frosts, that yields under the too adventurous skater, it could not support great weights,—table-lands such as now exist, or mountain chains; and hence, apparently, the existence of vast swampy plains nearly level with the sea, and ever-recurring periods of subsidence, wherever a course of deposition had overloaded the surface. The yet further fact, that as we ascend into the middle and earlier Palæozoic periods, the traces of land become less and less frequent, until at length scarce a vestige of a terrestrial plant or animal occurs in entire formations, seems charged with a corroborative evidence. I shall not say that in these primeval periods

'A shoreless ocean tumbled round the globe,'

for the terrestrial plants of the Silurians show that land existed in even the earliest ages in which, so far as the geologist knows, vitality was associated with matter; but it would seem that only a few insulated parts of the earth's surface had got their heads above water at the time. The thin and partially-consolidated crust could not bear the load of great continents; nor were the 'mountains yet settled, nor the hills brought forth.' It would seem that not until the Carboniferous ages did there exist a period in which the slowly-ripening planet could exhibit any very considerable breadth of land; and even then it seems to have been a land consisting of immense flats, unvaried, mayhap, by a single hill, in which dreary swamps, inhabited by doleful creatures, spread out on every hand for hundreds and thousands of miles, and a gigantic and monstrous vegetation formed, as I have shown, the only prominent features of the scenery. Burnett held that the earth, previous to the Flood, was one vast plain, without hill or valley, and that Paradise itself, like the *blomen garten* of a wealthy Dutch burgomaster, was curiously laid out upon a flat. We would all greatly prefer the Paradise of Milton—

> 'A happy rural seat of various view;
> Grooves whose rich trees wept odorous gums and balm;
> Others whose fruits, burnish'd with golden rind,
> Hung amiable, Hesperian fables true,
> If true, here only, and of delicious taste:
> Betwixt them lawns, or level downs, and flocks
> Grazing the tender herb, were interposed,
> Or palmy hillock; or the flowery lap
> Of some irriguous valley spread her store,
> Flowers of all hue, and without thorn the rose:
> Another side, umbrageous grots and caves
> Of cool recess, o'er which the mantling vine
> Lays forth her purple grape, and gently creeps
> Luxuriant; meanwhile murmuring waters fall
> Down the slope hills, dispersed, or in a lake,
> That to the fringed bank with myrtle crown'd
> Her crystal mirror holds, unite their streams.'

It was during the times of the Coal Measures that Burnett would have found his idea of a perfect earth most nearly realized in at least general outline; but even he would scarce have deemed it a paradise. Its lands were lands in which, according to the Prophet, there 'could no man have dwelt, nor son of man passed through.' From some tall tree-top the eye would have wandered, without resting-place, over a wilderness of rank, unwholesome morass, dank with a sombre vegetation, that stretched on and away from the foreground to the distant horizon, and for hundreds and hundreds of leagues beyond; the woods themselves, tangled, and dank, and brown, would, according to the poet, have 'breathed a creeping horror o'er the frame;' the surface, even where most consolidated, would have exhibited its frequent ague-fits and earth-waves ; and, after some mightier earthquake, had billowed the landscape, dashing together the crests of tall trees and gigantic shrubs, there would be a roar, as of many waters, heard from the distant outskirts of the scene, and one long wall of breakers seen stretching along the line where earth and sky meet,—stretching inwards and travelling onwards with yet louder and louder roar,—Calamite and Ulodendron, Sigillaria and Tree-fern, disappearing amid the foam,—until at length all would be submerged, and only here and there a few Araucarian tops seen over a sea without visible shore. Such was the character, and such were the revolutions, of the land of the Carboniferous era,—a land that seems to have been called into being less for the sake of its own existence than for that of the existences of the future.

LECTURE SIXTH.

Remote Antiquity of the Old Red Sandstone—Suggestive of the vast Tracts of Time with which the Geologist has to deal—Its great Depth and Extent in Scotland and England—Peculiarity of its Scenery—Reflection on first discovering the Outline of a Fragment of the Asterolepis traced on one of its Rocks—Consists of Three Distinct Formations—Their Vegetable Organisms—The Caithness Flagstones: how formed—The Fauna of the Old Red Sandstone—The Pterichthys of the Upper or Newest Formation—The Cephalaspis of the Lower Formation —The Middle Formation the most abundant in Organic Remains—Destruction of Animal Life in the Formation sudden and violent—The Asterolepis and Coccosteus—The Silurian the Oldest of the Geologic Systems—That in which Animal and Vegetable Life had their earliest beginnings—The Theologians and Geologists on the Antiquity of the Globe—Extent of the Silurian System in Scotland—The Classic Scenery of the Country situated on it—Comparatively Poor in Animal and Vegetable Organisms—The Unfossiliferous Primary Rocks of Scotland—Its Highland Scenery formed of them—Description of Glencoe— Other Highland Scenery glanced at—Probable Depth of the Primary Stratified Rocks of Scotland—How deposited—Speculations of Philosophers regarding the Processes to which the Earth owes its present Form—The Author's Views on the subject.

I INCIDENTALLY mentioned, when describing the Oolitic productions of our country, that the shrubs and trees of this Secondary period grew, on what is now the east coast of Sutherland, in a soil which rested over rocks of Old Red Sandstone, and was composed mainly, like that of the county of Caithness in the present day, of the broken débris of this ancient system. We detect fragments of the Old Red flagstones still fast jammed among the petrified roots of old Oolitic trees; we find their water-rolled pebbles existing as a breccia, mixed up with the bones of huge Oolitic reptiles and the shells of extinct Oolitic molluscs; we even find some of its rounded masses incrusted over with the *corals* of the Oolite: the masses had existed in that remote age of the world as the same grey indurated blocks of stone that we find them now; and busy Madreporites, —Isastræa and Thamnastræa,—whose species have long since perished, built up their stony cells on the solid foundations which the masses furnished. Nay, within the close compressed folds of these flagstones lay their many various fossils,—glittering scale, and sharp spine, and cerebral buckler,—in exactly the same state of keeping as now; and had there been a geologist to take hammer in hand in that Oolitic period, when the spikes of the *Pinites Eiggensis* were green upon the living tree, and the *Equisetum columnare* waved its tall head to the breeze, he would have found in these stones the organisms of a time that would have seemed as remote then as it does in the present late age of the world. We may well apply to this incalculably ancient Old Red system what Wordsworth says of his old Cumberland beggar,—

> ' Him from my childhood have I known ; and then
> He was so old, he seems not older now.'

This glimpse, through the medium of the high antiquity of the Oolite, of an antiquity vastly higher still,—that of the Old Red Sandstone,—may well impress us with the enormous extent of those tracts in time with which the geological historian is called on to deal. There are some of the lesser planets that seem to the naked eye quite as distant as many of those fixed stars whose parallax the astronomer has failed to ascertain; but when they come into a state of juxtaposition, and the moveless star is seen dimly through the atmosphere of the moving planet, we are taught how enormous must be those tracts of space which intervene between them, and keep them apart. And it is thus with the periods of the geologist. Even the comparatively near are so distant, that the remote seem scarce more so; but the dead and stony antiquity of one system, seen as if through the living nature of another, enables us, in at least some degree, to appreciate the vastness of those cycles by which they were separated. It is further interesting, too, thus to find one antiquity curiously inlaid, as it were, in another. We feel as if, amid the ancient relics of a Pompeii or a Herculaneum, we had stumbled on the cabinet of some Roman antiquary, filled with bronze and granite memorials of the first Pharaohs, or of the old hunter king who founded Nineveh;—things that in times which we now deem ancient had been treasured up as already grown venerably old.

The Old Red Sandstone underlies the Coal Measures, and is, in Scotland at least, still more largely developed than these, both in depth and lateral extent. In Caithness and Orkney, one of the three great formations of which it consists has attained to a thickness that equals the height of our highest hills over the sea.[1] The depth of the entire system in England has been estimated by Sir Roderick Murchison at ten thousand feet; and as these ten thousand feet include three formations so distinct in their groups of animal life that not a species of fish has been found common to both higher and lower, it must represent in the history of the globe an enormously protracted period of time.

The scenery of the Old Red Sandstone we find much affected to the south of the Grampians, like that of the Coal Measures, by the presence of the trap rocks; but in the north, where there is no trap, it bears a character decidedly its own. It is remarkable for rectilinear ridges elongated for miles, that, when they occur in semi-Highland districts, where the primary rocks have been heaved into wave-like hills, or ascend into boldly-contoured mountains, constitute a feature noticeable for the contrast which it forms to all the other features of the scene. In approaching the eastern coast

[1] The Caithness flagstones and their ichthyolites constitute, according to Sir R. Murchison, the *central* portion of the Old Red group.— W. S. S.

of Caithness from the south, the voyager first sees a mountain country,—the land piled up stern and high,—the undulations bold and abrupt. He is looking on the Highlands of Sutherlandshire. All at once, however, the aspect of the landscape changes;—the broken and wavy line suddenly descends to a comparatively low level, and, wholly altering its character, stretches away to the north, straight as a tightened cord, or as if described by a ruler. Caithness thus seen in profile reminds one of a long thin proboscis, or mesmerized arm, stretched stiffly out from the Highlands to the distant Orkneys. In sailing upwards along the Moray Firth, the line which defines seawards the plain of Easter Ross from the Hill of Nigg to the low rocky promontory of Tarbat, topped by its lighthouse, presents nearly the same rectilinear character. Another long straight line which meets the eye on entering the bay of Cromarty stretches westwards from the hill of granitic gneiss immediately over the town, and runs for many miles into the interior along the bleak ridge of the Black Isle. Yet another rectilinear line may be seen running on the south side of the Moray Firth, from beyond the Moor of Culloden, which it includes, to the eastern end of Loch Ness. And in all these instances the rectilinear ridges are composed of Old Red Sandstone. On some localities on the seaboard of the country the system is much traversed by firths and bays, and what in Caithness and Orkney are termed *goes*,—narrow inlets in the line of faults, along which the waves find straight passage far into the interior. From the Hill of Nigg, the centre of an Old Red Sandstone district, the eye at once commands three noble firths, all scooped out of the deposit,—the Firth of Cromarty, the Dornoch Firth, and the upper reaches of the Moray Firth. It commands, too, what is scarce less a feature of the Old Red system,—the rich corn-bearing plains of Moray and of Easter Ross; and from the union which the prospect exhibits of two elements dissociated elsewhere in the country,—the rich softness of a Lowland scene, with numerous arms of the sea, characteristic elsewhere, as on the western coast, of a Highland one, —it forms a landscape unique among the landscapes of Scotland. But perhaps the most striking scenic peculiarities of the Old Red Sandstone are to be found in its rock-pieces. The Old Man of Hoy, with its mural rampart of precipices, not unfurnished with turret and tower, and wide yawning portals, and that rise a thousand feet over the waves,—the tall stacks of Canisbay, ornately Gothic in their style of ornament, with the dizzy chasms of the neighbouring headland, in which the tides of the Pentland Firth for ever eddy and boil, and the surf for ever roars,—and the strangely fractured precipices of Holburn Head, where, through dark crevice and giddy chasm, the gleam of the sun may be seen reflected far below on the green depths of the sea, and, venerable and grey, like some vast cathedral, a dissevered fragment of the coast descried rising beyond,—are all rock-scenes of the Old Red Sandstone. When I last stood on the heights of Holburn, there was a heavy surf toiling far below along the base of the overhanging wall of cliff which lines the coast; and deep under my feet I could hear a muffled roaring amid the long corridor-like caves into which the headland is hollowed, and which, opening to the light and air far inland by narrow vents and chasms, send up at such seasons, high over the blighted sward, clouds of impalpable spray, that resemble the smoke of great chimneys. As I peered into one of these profound gulfs, and dimly marked, hundreds of feet below, the upward dash of the foam, grey in the gloom,—as I looked, and experienced, with the gaze, that mingled emotion natural amid such scenes which Burke so well analyses as a consciousness of great expansiveness and dimension, associated with a sense of danger,—my eye caught, on the verge of the precipce, the outline of part of an old reptile fish traced on the rock. It was the cranial buckler of one of the hugest ganoids of the Old Red Sandstone,—the *Asterolepis*. And there it lay, as it had been deposited, far back in the bypast eternity, at the bottom of a muddy sea. But the mud existed now as a dense grey rock, hard as iron; and what had been the bottom of a Palæozoic sea had become the edge of a dizzy precipice, elevated more than a hundred yards over the surf. The world must have been a very different world, I said, when that creature lived, from what it is now. There could have been no such precipices then; a few flat islands comprised, in all probability, the whole dry land of the globe; and that emotion of which I have just been conscious, is it not something new in creation also? The deep gloom of these perilous gulfs,—these incessant roarings,—these dizzy precipices,—the sublime roll of these huge waves,—are they not associated in my mind with a certain dim idea of danger,—a feeling of incipient terror, which, in all God's creation, man, and man only, is organized to experience? Is it not an emotion which neither the inferior animals on the one hand, nor the higher spiritual existences on the other, can in the least feel,—an emotion dependent on the union of a living soul with a fragile body of clay, easily broken?

The Old Red Sandstone consists, as I have said, of three great formations, furnished each, in Scotland at least, with its peculiar group of fossils. In the *upper* division,—that which rests immediately under the Carboniferous system,—a few straggling plants of the Coal Measures have been occasionally found; but, so far as I know, no plant peculiar to itself. In the middle (*lower*) division we find traces of a peculiar but very meagre flora. I detected about ten years ago, among the grey micaceous sandstones of Forfarshire, a fucoid furnished with a thick, squat stem, that branches into numerous divergent leaflets or fronds of a slim, grass-like form, and which, as a whole, somewhat resembles the scourge of cords attached to a handle with which a boy whips his top. And Professor Fleming describes a still more remarkable vegetable organism of the same formation, which, to employ his own well-selected words, 'occurs in the form of circular flat patches, composed each of numerous smaller contiguous circular pieces, altogether not unlike what might be expected to result from a compressed berry, such as the bramble or rasp.'[1] In the lowest (*middle*) division of the Old Red traces of land plants become very rare. Many years ago, at Cromarty, I detected, in one of its oldest fossiliferous beds, a fragment of a cone-bearing tree, remarkable as being the oldest piece of wood ever found, that, when subjected to the microscope, exhibits the true ligneous structure; and I possess a small specimen from Skaill, in the mainland of Orkney, also detected in one of the lower beds, which formed apparently a portion of some nameless fern; but the other vegetable remains of the lower (*middle*) division, though sufficiently abundant in some localities to give a fissile character to the rock in which they occur, are, with one doubtful exception, all marine. They were the weeds of a widely extended sea, in which land was at once very unfrequent and of very limited extent. In the neighbourhood of Thurso my attention has been attracted for several years past by a curious appearance among the flagstones of the district,—there enormously developed,—which I am disposed to regard as indications of the existence of vast mud flats of the Old Red Sandstone, that occasionally showed their surfaces above water for perhaps weeks and months at a time, but which were in every instance submerged ere they acquired coverings of terrestrial vegetation. The flagstones, now known very extensively over Europe as the Caithness flag of commerce, must have been deposited at the bottom of a shallow sea,

---

[1] *Parka decipiens.* See *Testimony of the Rocks*, latest edition. For notice of a Lepidodendron occuring in the Forfarshire sandstone see likewise *Testimony of the Rocks*, page 446-7.—L. M.

in the form of beds of arenaceous mud, largely charged with organic matter. They abound in minute ripple-markings, which could have been formed only a few feet, or at most a few fathoms, under the surface; and between these rippled bands, for a series of beds together, there occur bands which had been evidently subjected to a drying process, so that, as happens with the bottom of a muddy pool laid dry during the summer droughts, they cracked into irregularly polygonal divisions, and as, when again submerged, a sudden deposition filled up the cracks, we can still trace these marks of desiccation as distinctly in the stone as if they had been made by the sun of the previous week. Hall of Leicester spoke, in one of his illustrations, of 'a continent of mud;' and it would seem that in the earlier ages of the Old Red Sandstone continents of mud were not mere figures of speech; but that over dark-hued and shallow seas, mud-banks of vast extent occasionally raised their flat dingy backs, and remained hardening in the hot sun until their oozy surfaces had cracked and warped, and become hard as the sun-baked brick of eastern countries; and that then, ere the seeds of terrestrial plants, floated from some distant island, or wafted in the air, had found time to strike root into the crevices of the soil, some of the frequent earth-tremors of the age shook the flat expanse under the water out of which it had arisen, and the waves rippled over it as before. The features of a scene so tame and unattractive, —features which none of the poets, save perhaps the truthful Crabbe, would have ventured to portray,—will not strike you as very worthy of preservation: there is certainly not much to excite or gratify the fancy in a scene of wide yet shallow seas, here and there darkened by forests of algæ, and here and there cumbered by archipelagos of flat verdure-less islands of mud that harden in the sun; but, regarded as embryo and rudimentary land, even these mud-banks may be found to possess their modicum of interest. And we know that in the shallows of that muddy sea, the Creator wrought with all His wonted wisdom and inexhaustible fertility of resource, in the production of a dynasty of fishes of very extraordinary form, but high type, and which manifested exquisite faculties of adaptation to the circumstances in which they were placed.

In glancing at the fauna of the Old Red Sandstone, let us imagine three great platforms from which the sea has just retired, leaving them strewn over with its spoils,—chiefly fishes. These platforms represent the three great periods of the system; and in each do we find the group specifically, and in several instances generically, distinct. In the upper or newer platform,—that immediately under the Coal Measures,—there occur several species of Holoptychius, all of them of smaller dimensions than the giant of the Carboniferous system, but, in proportion to their bulk and size, even *more* strongly armed. With the Holoptychius there was associated a fish of the same Cœlacanth family, the Bathriolepis, and several curious fishes of what is known as the Dipterian family, such as the Stagonolepis[1] and Glyptolepis. It contains also at least three species of Pterichthys. One of these, the *Pterichthys major*, which occurs in the upper sandstones of Moray, is of greater size than any of the others its contemporaries, or than any of the older species; as if, in at least point of bulk, the creature received its fullest development just when on the eve of passing away.[2]

[1] The Stagonolepis is now under examination as to whether it is to be ranked as fish or reptile. Sir R. Murchison mentions this in his last address to the Leeds British Association as still undetermined.—L. M.

[2] Associated with this large Pterichthys are now found not only the Telerpeton Elginense, a small tortoise, but footprints of larger reptiles, some only of greater size than the Telerpeton, others considered to approach more nearly in bulk and conformation to some of those of the succeeding eras. When I lately visited the Museum at Elgin, I was gratified by seeing sandstone slabs bearing the traces of each of these; but I was told that the best specimens had been sent to London

This strange Pterichthyan genus first appears at the base of the Old Red Sandstone, and disappears with its upper beds. It is peculiarly and characteristically the distinctive organism of the system, for in no other system does it occur; and it has a yet further claim on our notice here, from the extreme singularity of its construction. 'It is impossible,' says Agassiz, in his great work on fossil fishes,—'it is impossible to find anything more eccentric in the whole creation than this genus. The same astonishment which Cuvier felt on examining for the first time the Plesiosauri, I myself experienced when Mr. Hugh Miller, the original discoverer of these fossils, showed me the specimens which he had collected from the Old Red Sandstone of Cromarty.' And we find Humboldt referring, in his *Cosmos*, to this strange Pterichthyan genus in nearly the same terms. This, I suspect, is no place for strict anatomical demonstration; and so, instead of elaborately describing the Pterichthys, I shall merely attempt sketching its general outlines by the aid of a few simple illustrations. When, in laying open the rock in which it lies, the under part is presented, as usually happens, we are struck with its resemblance to a human figure, with the arms expanded, as in the act of swimming, and the legs transformed, as in the ordinary figures of the mermaid, into a tapering tail. On further examination, we ascertain that the creature was cased in a complete armature of solid bone, but that the armour was of different construction over the different parts. The head was covered by a strong helmet, perforated in front by two circular holes, through which the eyes looked out. The chest and back were protected by a curiously constructed cuirass, formed of plates; and the tail sheathed in a flexible mail of osseous scales. The arms, which were also covered with plates, were articu-

for examination. It is probable that they will have been lawfully named and surnamed by the *savants* ere the next edition of this work is ready for the press.—L. M.

lated rather to the lower part of the head than to the shoulders; and this by what at first appears to be simply a ball-and-socket joint, like that of the human thigh, but which, on further examination, proves to be of a more complex character, as we find a pin-like protuberance from the socket finding, in turn, a socket in the ball. The abdomen of the creature was flat; the dorsal portions strongly arched; and not in our Gothic roofs, constructed on strictly mathematical principles, do we discover more admirable contrivances for combining in the greatest degree lightness with strength, than in the arch of osseous plates which protected the Pterichthys. Nay, we find in it the two leading peculiarities of the Gothic roof anticipated,—the contrivance of a series of ribs that radiate from certain centres, and the contrivance of the groin. The helmet was united to the cuirass by a curious and yet very simple joining, that united the principle of the dovetail of the carpenter to that of the keystone of the architect. Further, the creature, with its inflexible cuirass and its flexible tail, and with its two arms, that combined the broad blade of the paddle with the sharp point of the spear, might be regarded, when in motion, as a little subaqueous boat, mounted on two oars and a scull. And such was the Pterichthys,—the characteristic organism of the Old Red Sandstone. I may remark, in connexion with this fish,—a remark, however, which bears equally on all its ganoidal contemporaries,—that the development of its dermal or skin-skeleton, compared with that of its internal one, was singularly great. In the present creation, with but a few exceptions, such as the Pangolin and Armadillo among quadrupeds, the crocodiles and tortoises among reptiles, and the Lepidosteus and Polyopterus among fishes, the dermal skeleton is but very slenderly represented. In our own species, for instance, it is represented by but the teeth, the hair, and the nails; and were there no other portions of us to survive in the fossil state, each of the male animals among us would be represented by but ten toe and ten finger nails, one set of teeth, a periwig, and a pair of whiskers. But so complete, on the other hand, was the development of the dermal skeleton among the fishes of the Old Red Sandstone, that, though in many instances no other parts of them survive, we find their outlines complete in the rock from head to tail. Dermal plates of enamelled bone represent the head; dermal scales, also of enamelled bone, lie ranged side by side, like tiles on a roof, in the lines in which they originally covered the body; and thickly-set enamelled rays of bone indicate the place and outline of the fins. As a set-off, however, against this great development of dermal skeleton in the ganoids of the Old Red Sandstone, their internal skeletons were exceedingly slight, and in whole families entirely cartilaginous.

The middle (*lower*) platform of the Old Red Sandstone has for its characteristic organism the Cephalaspis, or Buckler-head,—a curiously formed, bone-covered fish, with a thin triangular body, and crescent-shaped head, somewhat resembling in outline a shoemaker's cutting-knife. It had for its contemporaries several fishes armed with dorsal spines, of which only the spines remain, and of a gigantic Crustacean, akin, as shown by some of its plates, to our existing lobsters, but which in some specimens must have exceeded four feet in length.

It is, however, on the lower (*middle*) platform of the system that we find its organic remains at once most abundant and most characteristic. The flagstones of Caithness and Orkney, and the nodule-bearing beds of Ross, Cromarty, and Moray, contain more fossil fish than all the other formations of not only Scotland, but of Great Britain, from the Tertiary deposits down to the Mountain Limestone. There are strata in which they lie as thickly as herrings on our better fishing banks in autumn, when the fisherman's harvest is at its best; and, strange to say, not unfrequently do the fish of a whole platform give evidence, both in their state of keeping and in their contorted attitude, that they all died at once, and died by violent death. We see them still presenting over wide areas the stiff curved outline,—a result of the unequal contraction of the muscles,—which, as in the case of recently netted herrings, marks that dissolution had been sudden. We find, too, that their remains did not suffer from the predatory attacks of other fishes: it would seem as if all the finny inhabitants of wide tracts of sea had been at once cast dead to the bottom, so that not an individual survived, to prey upon the remains of his deceased neighbours. It was the first remark of Agassiz, when introduced to a collection of fossil fish from Orkney,—'All these fish died by violent death,'—a remark which he again and yet again repeated when introduced to the Old Red ichthyolites of Cromarty and Moray. We have already seen that the oldest plant-covered land of which the geologist finds distinct and certain trace in this country was a land subject to incessant fluctuations of level, and to sudden and disastrous invasions of the sea; and that, though suited for the production of a rank and luxuriant flora, whose numerous denizens lived without consciousness and died without suffering, or for animals fitted to enjoy the present without thought or fear of the future, and to whom life, so long as they lived, was pleasure, and death merely a ceasing to be, we conclude that it could have been no fitting home for creatures of a higher order, whose nature it is to look before and behind them,—before them with hope or with fear, behind them with satisfaction or regret. And these strange platforms of sudden death,—of no rare occurrence in the marine depths of the Old Red Sandstone,—show that the sea in these early times was not less subject to disastrous catastrophe than the land,—that that order of nature which we now term its fixed order, and on whose permanency our minds have been framed to calculate, was, if I may venture the expression, enacted, but not enforced, and so the breaches of it were scarce more exceptional than the observance,—that life, greatly more emphatically than now, was the least certain of all things,—and that both in sea and on the land the young and immature earth, like an inexperienced and careless nurse, was ever and anon overlaying and smothering its offspring.

Among the various ichthyic families and genera of the Lower and Middle Old Red Sandstone,—Acanths, Dipterians, Cœlacanths, and Cephalaspians,—I shall refer to only two, and that in but a few brief words; the one remarkable for its great size, the other for its extraordinary organization. The *Asterolepis* seems to have been one of at once the earliest and bulkiest of the ganoids. Cranial bucklers of this creature have been found in the flagstones of Caithness large enough to cover the front skull of an elephant, and strong enough to have sent back a musket-bullet as if from a stone wall. The Asterolepis must have at least equalled in size the largest alligators; and there were several points in which it must have resembled that genus of reptiles. Its head was covered with strong osseous plates, ornately fretted by star-like markings, and its body by closely imbricated and delicately-carved osseous scales. But it is chiefly in its jaws that we trace a reptilian relationship to the alligators. The alligators among existing reptiles, and the Lepidostei among existing reptile-fishes, are remarkable for a peculiar organization of tooth and maxillary, through which certain long teeth in the anterior part of the nether jaw are received into certain scabbard-like hollows in the anterior part of the upper jaw. The hollows receive the teeth when the mouth is shut, as the scabbard receives the sword. Now, in the Asterolepis this reptilian peculiarity was not restricted to a small group of the anterior teeth, but pervaded the entire jaw. Beside each of the creature's reptile teeth, in both jaws, there was a deep pit, which received the reptile tooth opposite; and thus, when the animal closed its formidable mouth, the jaws would have been locked together by their long teeth and deep recipient hollows, as the crenellated jaws of a fox-trap lock into each other when we release the spring. The other ichthyolite of the Old Red Sandstone to which I shall refer is the *Coccosteus*,—a ganoid that, so far as we yet know, was restricted to this formation. Like the Pterichthys, with which it has been classed, it was provided with a helmet and cuirass of bony plate; but its caudal portion seems to have been naked,—a peculiarity of which we find no other example among the ganoids of this early time. The Coccosteus was, however, chiefly remarkable for the form of its jaws.[1] More than ten years ago I ventured to state, in the first edition of a little work on the Old Red Sandstone, that the jaws of this ancient fish seemed, like those of some of the crustaceans, and of some of the insects, to have possessed a horizontal action. Aware, however, that I was on dangerous ground, I exercised, in making the statement, some little share of Scotch caution: the thing was, I stated, too anomalous to be regarded as proven by the evidence of the specimens yet found; and I mentioned it, I said, only with the view of directing attention to it. It was a question, I thought, worthy of being entertained, and so I craved that it *should* be entertained, and specimens carefully examined. But specimens were *not* examined, at least no specimens that threw any light on the subject; and my very modified statement respecting it was written down a blunder on the very highest authority. I kept, however, a steady eye on the rocks, as the real authorities in the case; and, deeming myself bound as a geologist to observe carefully and record truthfully whatever they revealed, but as not in the least responsible for the anomalies of the revelation, I persisted in quietly collecting their evidence in a

[1] The Coccosteus possessed also true bony vertebræ. See *Siluria*, p. 504, new edition.—W. S. S.

suite of fossils, which has now fully convinced our first comparative anatomists that there was an anomaly in the structure of the jaws of this ancient fish, unique among the vertebrata; and that, in calling to it the attention of the scientific world, I was in the right, not in the wrong. The under jaws contained two distinct sets of teeth; the one set or group in the line of the symphysis, the other set or group on the upper edge of the jaw, and placed on such different planes, that they could not possibly have been brought into action by the same movement of the condyles. And there are on the table specimens which show, that while the group in the customary place, the upper edge of the under jaw, were made to act against a group placed in the nether edge of the upper one by the usual vertical action, the groups so strangely placed in the symphysis, if brought into action at all, must have acted against each other through a lateral motion altogether unique. The jaws of the Coccosteus are interesting in another point of view, as being perhaps the oldest portions of any internal skeleton that have presented their structure to the microscope. And it is surely not uninteresting to see the osseous substance, destined to perform so important a part in the animal economy, presenting in so early an age its distinguishing characteristics; in especial, those arterial Haversian canals through which the ancient blood must have flowed for its nourishment, and those numerous corpuscles or life-points from which its organization began, and which continued to remain open as the sheltering cells in which its vitality resided. Was it impossible, in the nature of things, we ask, that life could be equally diffused over hard and rigid earth built up into this new animal substance, bone? and was it therefore merely sown over it in hollow microscopic points? Is bone rather a thing strongly garrisoned by vitality, than itself vital? Direct questions cannot always, in the present imperfect state of our knowledge, receive answers equally direct; and these are questions to which our first physiologists might hesitate to reply. But we may at least safely infer, from the thorough identity of the osseous material throughout all ages, that it was a material compounded at all times by the same Architect, according to a predetermined recipe; that it is He who built up the corpuscles and arranged the canals in that ancient jaw which so excites our curiosity, that now maketh in the human subject 'the bones to grow;' and that, in his eternal purposes, the existences of the most ancient times may be woven into the tissue of one great plan, with all that now exists, and with all that shall exist in the future.

In retiring into the remote past, and descending from formation to formation as we retire, we have now reached that great Silurian group of rocks in which, so far as the geologist yet knows, fossils first appear, and which represents a period of incalculable vastness, in which life, animal and vegetable, seems to have had its earliest beginnings on our planet. Enormous as is the depth of some of the other systems,—such as the Old Red and the Carboniferous systems,—they shrink into moderate dimensions when we compare them with the truly vast Silurian deposit. It was estimated only a few years ago, that the entire depth of all the fossiliferous strata did not much exceed six miles: it is now found by the geologists of the Government Survey, that the Lower Silurian strata of North Wales are of themselves about five miles in depth, while the Upper Silurian, as estimated by Sir Roderick Murchison, are about a mile more. Many of the beds, too, of both the Upper and Lower divisions, must have been of exceedingly slow deposition,—formed far from land, and at the bottom of deep seas: nay, there are Silurian limestones that can scarce be regarded as deposits at all, seeing that every calcareous particle of which they are composed was at one time associated with animal life, as the joints of crinoidea, the calcareous framework of corals, or the shells of molluscs, all of which lived and died upon the spot that the rocks now occupy. And rocks of this character, when of any considerable thickness, must have been very many years in the forming. The sagacious Chalmers saw and taught, at the beginning of the present century, that 'the writings of Moses do not fix the antiquity of the globe :' 'if they fix anything at all,' he said, 'it is only the antiquity of the species.' But there were few among either teachers or pupils who saw so clearly as Chalmers; and when the geologist first began to demand a long tale of years for the production of all the stony volumes of his record, it was, like the long price which the ancient sibyl demanded for all *her* volumes, very decidedly refused him. Instead, however, of bating in the demand, or acquiescing in the denial, the geologists have been ever and anon returning, sibyl-like, to drive harder and yet harder bargains, and even to ask, as they do now, as much for a single volume as they formerly asked for the whole ; but their library, unlike that offered in sale to the old Roman, is undergoing no diminution in bulk ; on the contrary, its volumes increase in number as the demand made for each is raised. But it is at least something to be made to feel, by means of these time-marks in the remote distance, that eternity is not a mere idle name, which at times children employ in their catechisms, but a great and awful fact; and that its unmeasurable amplitude of duration closes as completely around the systems of the geologist in time, as the infinity of extension closes around the systems of the astronomer in space. It is one of the revealed characteristics of the adorable Creator, that 'from everlasting to everlasting he is God.'

On the western coast of Ross and Sutherland, on a general basement of broken primary hills of no great altitude, we find the (*Cambrian*) deposit occurring as a series of noble mountains, now entirely insulated from each other, and that yet give evidence, in their lines of nearly horizontal strata, that they once formed parts of a continuous bed, which ere the operation of the denuding agencies, had overlaid, to the depth of from two to three thousand feet, the gneiss and quartz deposits below. They now exist, however, as a group of magnificent pyramids, compared with which those of Egypt are but the toy erections of children; and yet, from the rectilinear character of their abrupt and mural precipices, coursed as if with tiers of ashlar,—from their general regularity of form, their utter bareness of vegetation, and their rich warm colour, which contrasts as strongly with the cold grey tints of the rocky platform on which they rest, as the warm colour of our fresher public buildings with the cold grey of our paved streets or squares,—they seem rather works of human contrivance than productions of Nature. Seen from the west in a clear summer evening, when the red level light falls on the still redder stone, but at a sufficient distance to admit of those softening influences of the atmosphere which mellow the harsher reds into crimson and purple, there is a gorgeous beauty in these pieces of Nature's masonry which it is scarce possible to exaggerate in description. Beneath and in front we see a tumbling sea of craggy hills, which even the warm gleam of sunset scarce relieves from their sober tint of neutral grey; while rising over them abrupt and bold, and lined with their horizontal bars, appear the noble pyramids in their rich vestures of regal purple,—the monuments of an antiquity compared with which that of Nineveh and Babylon belong to the morning hours of a day not yet come to its close.[1]

But it is peculiarly in the southern Silurian portions of the kingdom that 'scarce a mountain lifts its head unsung.'

[1] The above description of the scenery of the West Highlands is, in fact, that of the Silurian, although written before Sir Roderick Murchison discovered his error in laying down these mountains as Old Red. It is inserted here to fill up the hiatus in description which would else occur.—L. M.

Yarrow, Ettrick, St. Mary's Loch, Leader Haughs, Tweedside,—especially along those upper reaches of the river where it mirrors, in its calmer pools, the classic ruins of Melrose and Dryburgh, and the young woods of Abbotsford,—the Gala-water, Teviotdale, Lammermuir, Galloway, and Nithsdale, the springs of the Doon, the hills that rise over the source of Dee, and the 'moors and mosses many' where the 'Stinchar flows,'—are all to be sought and found in the Silurian region of Scotland. It will scarce do now to estimate the scenic merit associated with these names at its actual value. The words of sober truth would seem, according to Wordsworth, 'strange words of slight and scorn,'—

> 'What's Yarrow but a river bare,
>   That glides the dark hills under?
> There are a thousand such elsewhere,
>   As worthy of our wonder.'

Even the indomitable good-nature of Sir Walter was scarce proof against what he deemed the disparaging, but, I doubt not, truthful, estimate of Washington Irving. 'Our ramble,' says this accomplished writer, in his *Abbotsford*, 'took us on the hills, commanding an extensive prospect. "Now," said Scott, "I have brought you, like the pilgrim in the *Pilgrim's Progress*, to the top of the Delectable Mountains, that I may show you all the goodly regions hereabouts. Yonder is Lammermuir and Smailholme; and there you have Galashiels, and Torwoodlee, and Gala-water; and in that direction you see Teviotdale and the braes of Yarrow, and Ettrick stream winding along, like a silver thread, to throw itself into the Tweed." He went on thus to call over names celebrated in Scottish song, and most of which had recently received a romantic interest from his own pen. In fact, I saw a great part of the border country spread out before me, and could trace the scenes of those poems and romances which had in a manner bewitched the world. I gazed about me for a time with mute surprise, I may almost say with disappointment. I beheld a mere succession of grey, waving hills, line beyond line, as far as my eye could reach, monotonous in their aspect, and so destitute of trees, that one could almost see a stout fly walking along their profile; and the far-famed Tweed appeared a naked stream, flowing between bare hills, without a tree or thicket on its banks. I could not help giving utterance to my thoughts. Scott hummed for a moment to himself, and looked grave. He had no idea of having his muse complimented at the expense of his native hills. "It may be partiality," said he at length; "but to my eye these grey hills, and all this wild border country, have beauties peculiar to themselves. I like the very nakedness of the land: it has something bold, and stern, and solitary about it."' Yes; there is no question that, had not the poets thought so, they could not have sung so honestly and warmly, and, of consequence, so successfully:

> ' The poet's lyre, to fix his theme,
> Must be the poet's heart;'

and so let us with a good grace acquiesce in their decision. The border land, with its Silurian groundwork, has its peculiar beauties; and no one could portray them at once so graphically and so discriminately as Scott himself. Take, for instance, the passage in *Guy Mannering* where he describes his hero, Brown, and the redoubtable Dandie Dinmont, approaching Charlieshope after the rencontre with the gipsies on Bewcastle Moor :—' Night was now falling, when they came in sight of a pretty river winding its way through a pastoral country. The hills were greener and more abrupt than those which Brown had lately passed, sinking their grassy sides at once upon the stream. They had no pretensions to magnificence of height or to romantic shapes, nor did their smooth swelling slopes exhibit either rocks or woods; yet the view was wild, solitary, and pleasingly rural. No enclosures, no roads, almost no tillage: it seemed a land where a patriarch would have chosen to feed his flocks and herds.' This is faithful description, at once beautiful and characteristic; and such of my audience as remember the exquisite landscape of the 'Enterkin' of our countryman Harvey, as exhibited at the Royal Institution here, in, if I remember aright, the year 1846, with its grey rocks, its green swelling hills of softish outline, and its recluse and houseless valley of deepest loneliness, will be convinced, as I am, that where there is in the mind a certain prominent requisite present, the region of the Silurians is as available for the purposes of the painter as for those of the poet,—that one requisite being the not very definable and many-sided faculty represented by the single magic word *genius*.

The Silurians of Scotland, though of very considerable depth, are greatly less rich in organic remains than the contemporary deposits of England and the Continent. Vast beds of grey slaty rock, hundreds of feet in thickness, seem to have been formed at the bottom of profound seas beyond the zero line of animal or vegetable life. And even in the cases in which organisms of both kingdoms were present, we find their remains very imperfectly preserved.[1] The flora of the system in Scotland is represented merely by a few dark-coloured carbonaceous beds, which occasionally pass into an impure anthracite or blind coal, and which are probably identical in their origin with the anthracite schists of Scandinavia, regarded by Sir Roderick Murchison as the remains of large forests of algæ and fuci, which originally existed in the

---

[1] As mentioned in the preface, it is stated by Sir Roderick Murchison, in his Leeds address to the British Association, that twenty species of Silurian fossils have been discovered by Mr. Peach in a limestone band above the Silurian conglomerate of the Western Highlands, determined by Mr. Salter, and carefully examined by Sir Roderick himself. They are Maclurea, Murchisonia, Cephileta, and Orthoceras, with an Orthis, etc.—L. M.

Silurian seas, and which, from their perishable nature, have lost all trace of their original forms. In the ashes of an anthracite of our Scottish Silurians which occurs near Traquair, Professor Nicol of Cork observed, under the microscope, tubular fibres unquestionably vegetable, but which he thought indicative of vegetation of a higher class than our existing algæ. There is, however, a family of marine plants, now represented on our coasts by a single species, which had, I am inclined to think, its representatives at a very early period in our seas; and which, had it existed during the Silurian ages, could have furnished the tubular cells. I refer to the Zostera, or grass-wrack, a plant of the pond-weed family, which, unlike any of the algæ, has true roots, true flowers, true seeds, tough fibrous stems, and grass-like leaves, traversed by parallel veins, and that yet lives in the sea among *laminariæ* and *floridiæ*, far below the fall of our lowest streamtides. It is worthy of notice, too, that the *Zostera marina*, our recent British species, when driven ashore on parts of our coasts at certain seasons, as it always is in 'great abundance, decomposes into a substance much resembling peat, that, unlike the brown pulpy mass into which the algæ in similar circumstances resolve, retains distinct trace of the vegetable fibre. It is further noticeable, that some of the vegetable remains of the Old Red Sandstone,—the oldest specimens furnished by our Scottish flora that present aught approaching distinctness of outline,—exhibit several traits that remind us of the leaves of gigantic Zostera. The vegetable impressions of some of the Caithness flagstones have rectilinear edges, and are traversed by parallel lines, scarce less strongly marked than the ridges of the Calamite; but, from the extreme thinness of the impression left in the rock, they seem rather the veins of leaves than the fluted markings of stems. It is quite possible, therefore, that though the anthracite beds of our Scottish Silurian system give evidence of the existence of a higher vegetation than that of the algæ, it may have been a marine vegetation notwithstanding. No terrestrial plant has yet been detected in the Silurians of either England or Scotland : the flora of the time, within at least the area of the British islands, seems to have been a poverty-stricken flora of the sea, consisting mainly of Fuci and Algæ, and including as its highest forms a species or two of Zostera, or, as is more probable, of some extinct analogous family.

The Silurian fauna in Scotland consisted also, so far as we can now judge from the broken remains, of but a few marine forms. In the Silurian deposits of England fishes appear ; but in our Scotch Silurians we find nothing higher than a Trilobite or a cephalopodous mollusc. The Trilobite was perhaps the most characteristic organic form of the system. It occurred also, though in types specifically distinct, in the Old Red Sandstones of England and the Continent ; and I have found well-marked specimens even in the Mountain Limestone of this neighbourhood,—the formation in which the family finally disappears ; but it was in the Silurian system that it received its fullest development both in size and number ; and portions of at least five species have been detected in the Silurian deposits of Scotland. The Trilobite was a many-jointed crustacean, which, since the close of the Carboniferous period, has had no adequate representative in creation, but whose nearest allies we have now to seek among the minute Entomostraca, especially among the genus Branchipus,—little insect-like creatures, occasionally found in stagnant pools, furnished with fin-like legs, fitted-for swimming, but not for walking with, and that, spending happy lives, darting hither and thither through the upper reaches of the water, now swim along the surface on their backs and now on their abdomens. The Trilobites, like the Entomostraca, seem to have been furnished with merely membranaceous, oar-like limbs, and must have led a purely aquatic life as swimmers,—at one time oaring their way, back below, along the surface of the sea, at another, back above, along the bottom. But some of these Entomostraca of the old Silurian ocean were, compared with their modern representatives, of great size. The *Homalonotus delphinocephalus* had a carapace as large as that of an ordinary market-crab, and the *Asaphus tyrannus* and *Isoletus megistos* were each of them as large animals, though different in their proportions, as ordinary market-lobsters. But it seems to have been characteristic of both the flora and fauna of these ancient times, that many of their characteristic forms should unite great size to a humility of organization restricted in the present ages to forms comparatively minute. The Trilobites of the Silurian system, like the Club-mosses and Equisetaceæ of the Coal Measures, were of a Brobdingnagian cast; and, regarded as Entomostraca, we must hold—to return to a former illustration—that we look upon them with eyes sharpened by an experience acquired among the productions of Lilliput. So far as we yet know, the higher contemporaries of the Trilobite in Scotland were chambered shells of two well-marked genera,—that of the Orthoceratite, a long, straight, horn-shaped shell; and that of the Lituite, which may be described as an Orthoceratite curled up into a scroll. And, associated with these, we find some of the low brachiopodous molluscs of the more ancient types, such as Leptenæ, Orthes, and Spirifers. But by far the most characteristic organisms of our Scottish Silurians belonged to a low zoophytic family, allied by some of their affinities, in some of their genera, to the sea-pens, and by certain other affinities, in some of their other genera, to the Sertularia. They are known to the geologist by the general name of Graptolites. The Sertularia, compound, plant-like animals, that resemble miniature bushes in spring, just as the buds are bursting into leaf, are attached always, by their seeming roots, to rocks, shells, or sea-weed, and so require a hard bottom; whereas the sea-pens, compound, feather-like Zoophytes, whose every fibre contains its rows of living creatures, affect soft muddy bottoms, in which they may be found sticking by their quill-like points, like arrows in the soft sward around a target. I have seen them brought up by scores on the lines of the fisherman, out of a muddy ravine in the Moray Firth that sinks abruptly from beside the edge of a hard submarine bank, to the depth of thirty fathoms; and have often admired their graceful, quill-like forms, and their delicate hues, that range from pink to crimson, and from crimson to purple. And, judging from the character of those grey carbonaceous deposits in which the Graptolites of our Silurian rocks most abound, it is probable that they also were mud-loving animals, and more resembled in their habitats, if not in their structure, the sea-pens than the Sertularia. It is a curious circumstance that, in the group at least, the Graptolites of Scotland are more obviously allied to the Graptolites of the vast Silurian deposits of Canada and the United States, than to those of the Silurians of England. With this curious zoophyte we take farewell in Scotland of life and organization, and the record of the palæontologist closes. The remains of no plant or of no animal have been detected in this country underlying the rocks in which the oldest Graptolites occur.

Beneath the SILURIAN deposits of Scotland there rest, to an enormous thickness, what, with the elder geologists, I shall persist in terming the *primary deposits*, consisting, in the descending order, of clay-slates, mica-schists, quartz-rocks, primary limestones, and the two varieties of gneiss, —the granitic and the schistose.[1] In retaining the old name, I must, however, be regarded as merely holding that

[1] Hugh Miller evidently MORE THAN SUSPECTED the history of the geology of the north and north-west of Scotland, as developed by Mr. Peach and Sir Roderick Murchison in 1858.—W. S. S.

these rocks were actually the *first-formed rocks of what is now Scotland,—that the gneiss was gneiss, and the slate was slate, ere ever our oldest fossiliferous formations began to be deposited, or the organisms which they contain had lived or died.* Into the question raised regarding the form in which they were deposited, or the condition of our planet during the period of their deposition, I do not at present enter. On the other point, however,—the comparative antiquity of these unfossiliferous rocks in Scotland,—the evidence seems very conclusive: the base of some of the oldest deposits in which we find organisms enclosed consists of broken, and in most cases water-rolled, fragments of the gneisses, quartz-rocks, clay-slates, and mica-schists of the primary regions of the country.[1] These primary regions are of great extent. The *gneiss* region contains nearly ten thousand square miles of surface; the mica-schists, fully three thousand; and the quartz-rock and clay-slate united about fourteen hundred miles more. Comprising almost all the Highlands of Scotland, with the greater part of two of our Lowland counties, Banffshire and Aberdeen, their entire area, if we add about fifteen hundred miles additional of granite and primary porphyry, does not fall short of sixteen thousand square miles. It would be a bold and perilous task for one who has in some degree appreciated those sublimely impressive word-paintings of the Highlands which have added so largely to the well-earned celebrity of your distinguished President, and which seem invested with the very atmosphere of our hills, or who has seen with admiration and delight not only the very features, but all the poetry, of our noble mountain scenery, glowing from the canvas of Macculloch and of Hill,—it would, I say, be a perilous task, under the recollection of achievements such as theirs, to attempt a dull analysis of the geologic principles on which the peculiarities of our Highland

[1] See Murchison's *Siluria*, 2d edition, App. 553, 554, and 556.

landscape depend. I would feel as if I were bringing you from the studio of some heaven-taught sculptor, crowded with shapes of manly beauty and feminine loveliness, to lecture, amid the melancholy rubbish of a dissecting-room, on the articulations and proportions of the bones, and the form and position of the muscles. I shall venture, therefore, on merely a few desultory remarks, and shall request you, in order to lighten them as much as possible, to accompany me, first, in a sort of mesmeric expedition to the western extremity of Glencoe; at which, after having journeyed as only the clairvoyant can journey, let us now deem ourselves all safely arrived, and just set out on our way back again by the Loch Lomond road. In the course of our journey we shall pass, in the ascending order, over all the great Primary formations.[1]

Let us first mark the character of the Glen,—not less famous for the severe and terrible sublimity of its natural features, than for that dark incident in its history which associates in such melancholy harmony with the terrible and the severe. We are in a region of primary porphyry,—in the main a dark-coloured rock, though it is one of its peculiar traits, that in the course of a few yards it sometimes changes its hue from dark green to black or a deep neutral tint, and from these again to chocolate colour, to brick red, or to iron grey. But the prevailing hues are dingy and sombre; and hence, independently of the brown heath and ling, and those deep shadows which always

[1] According to a diagram which I have had the honour of receiving from the hand of Sir Roderick Murchison, illustrating his latest explorations in the north, there are two distinct gneisses,—an older and a younger; the first underlying the Cambrian conglomerate and Silurian fossil-bearing band of the west; the other or younger gneiss forming part of the central nucleus, and underlying the Old Red Sandstone conglomerates and ascending fossiliferous series of the east. Of course, the Cambrian will contain fragments of the older, and the Old Red conglomerate fragments of the younger gneiss.—L. M.

accompany steep rocks and narrow ravines, a sombre tone in the colouring of the landscape. When, however, for a few days the atmosphere has been dry and the sky serene, the dark rocks seem in many parts as if strewed over with an exceedingly slight covering of new-fallen snow,—the effect of the weathering of a thin film of the compact feldspar, which forms the basis of the porphyry into a white porcelanic earth. It is, however, in the form of the rocks that we detect the more striking peculiarities of the porphyritic formation. They betray their igneous origin in their semi-columnar structure. Every precipice is scarred vertically by the thick-set lines which define the thin irregular columns into which the whole is divided; and as the columnar arrangement is favourable to the production of tall steep precipices, deep narrow corries, and jagged and peaked summits, the precipices on either side are tall and steep, the corries are deep and narrow, and the summits are sharp, spine-like, and uneven. A hill of primary porphyry, where not too much pressed upon by its neighbour hills, as trees press upon one another in a thick wood, so that each checks the development of each, generally affects a pyramidal form; and we find fine specimens of the regularly pyramidal hill in the upper part of the valley, just as we enter on the open moor. I may mention, ere we quit Glencoe, that the more savagely sublime scenery of Scotland is almost all porphyritic. There is only one other rock,—hypersthene,—which at all equals the primary porphyry in this respect; and hypersthene is of comparatively rare occurrence in Scotland. It furnishes, however, one very noble scene in the Isle of Skye: the stern and solitary valley of Corriskin, so powerfully described in the *Lord of the Isles*, is a hypersthene valley.

Emerging from Glencoe, we enter upon a scene that, in simple outline, abstracted from the dingy tone of the colouring, and the bleak and scanty vegetation common to both, contrasts with it more strongly than perhaps any other in Scotland. We have quitted the porphyritic region, and entered upon a region of granite and gneiss. Looking back from that most solitary of Scottish inns, King's House, we find that we can determine with much exactness, from the form of the hills, where the porphyry ends and the granite or gneiss begins. The last of the porphyritic hills is a noble pyramid, broken into dizzy precipices, and lined vertically, like some of our semi-columnar traps; whereas the first of the granitic hills, placed immediately beside it, with but a narrow valley between, is of rounded outline,—a mere hummock magnified into a mountain, and wrapped round by a continuous caul of brown heath. On the other hand, we see the granite rolling out into a moory plain,—one of the dreariest in Scotland,—and forming a basin for a long, flat-shored loch, whose brown waters do not reflect a single human dwelling. Granite, however, does not always present features so little attractive. It is, in truth, a many-charactered rock. In general, the feldspar, which enters so largely into its composition, contains a considerable percentage of potash, and so decomposes readily; and hence the rounded forms of many of our granite hills and boulders. It affects, too, on the large scale, though unstratified, a tabular arrangement, and sometimes exists, as in this instance, and in those dreary parts of the lowlands of Aberdeen where the patrimony of the redoubtable Sir Dugald Dalgetty lay, as extensive and usually very barren plains. But in other parts it has little or no potash in its composition; and forming, in these circumstances, one of the most durable of rocks, its peaks and precipices stand up, as in Goatfell in Arran, with all the porphyritic sharpness of outline, unweathered for ages, or present, as in Ben Muich Dhui and its Titanic compeers, features at once bold, broad, and sublimely impressive. Humboldt, generally so correct, in his *Views of Nature*, seems to have seized on the granite in but one of its aspects. 'All formations,' we find him saying, 'are common to every quarter of the globe, and assume the like forms. Everywhere basalt rises in twin mountains and truncated cones; everywhere trap porphyry presents itself to the eye under the form of grotesquely-shaped masses of rock; while granite terminates in gently rounded summits.'

We pursue our journey, and enter on a great gneiss district. And in its swelling hills, rolled, like pieces of plain drapery, into but a few folds, and in its long withdrawing valleys, more imposing from an element of simple extent than from aught peculiarly striking in their contour, we recognise the staple scenery of the Scotch Highlands,—the scenery of ten thousand square miles. A gneiss hill is usually massive, rounded, broad of base, and withal somewhat squat, as if it were a mountain well begun, but interdicted somehow in the building, rather than a finished mountain. It seems almost always to lack the upper storeys and the pinnacles. It is, if I may so express myself, a hill of one heave; whereas all our more imposing Scottish hills,—such as Ben Nevis and Ben Lomond,—are hills of at least two heaves; and hence, in journeying through a gneiss district, there is a frequent feeling on the part of the traveller that the scenery is incomplete, but that a few hills, judiciously set down upon the tops of the other hills, would give it the proper finish. No hill, however, accomplishes more with a single heave than a gneiss one; the broad-based Ben Wyvis, that raises its head, white with other snows than those of age, more than three thousand feet over the sea, and looks down on all the other mountains of Ross-shire, is a characteristic gneiss hill of a single heave. Quitting the gneiss region, we cross a comparatively narrow strip of quartz rock. The quartz hills in its course are, however, not very characteristic. Such of you as may have sailed over the upper reaches of Loch Maree, with its precipitous, weather-bleached pyramidal hills, so bare of vegetation atop that their peaks

P

may be seen gleaming white in the autumnal moonlight for miles, as if covered with snow, or who may have threaded your way through the deep and sterile valleys that open their long vistas towards the head of the lake, will be better able to conceive, than from aught witnessed in the course of our present day's journey, of the savage wildness of scenery, —savage and wild, but grand withal,—which is the proper characteristic of a quartz-rock district.

And now, the strip of quartz rock passed over, we enter into an extensive region of mica-schist,—a formation so favourable to the development of a picturesque beauty,— ever and anon rising into the sublime,—that what is peculiarly the classic ground of Highland scenery is to be found within its precincts. Loch Awe, Loch Long, Loch Goil, Loch Tay, by much the larger and finer part of Loch Lomond, all Loch Katrine, Ben Venue, Ben Ledi, Ben Lomond, and the Trossachs, with many a fine lake and stream besides, and many a noble hill, are included in this rich province of the mica-schist.

We first become aware that we are nearing the formation, by the peculiar contour of its hills, as seen at a distance of several miles. As we approach their grey rocks of silky lustre, we find that they are curved, wrinkled, contorted, so as to remind us of pieces of ill-laid-by satin, that bear on their crushed surfaces the creases and crumplings of a thousand careless foldings; and mark further, that it is to these curves and contortions of the strata that the tubercled outlines of the hills are owing, and, with these, the bold projecting knobs and sudden recesses which break up their surfaces into so many picturesque wildernesses of light and shade. Not unfrequently, however, vast masses of schist, of a structure as dense and solid as that of granite, occur in the micaceous districts; and these form hills of a simpler outline, which, like the rock which composes them, seem intermediate in character between the mica-schist and the gneiss hills. All the mica-schists, however, decompose into soils, which, though light and thin, are more favourable to the production of the grasses and the common dicotyledonous shrubs and trees of the Highlands, than any of the gneisses or granites, and greatly more so than the porphyries or quartz rocks; and so the micaceous regions are not only more picturesque in outline than any of the others, but also richer in foliage and softer in colour. A tangled profusion of vegetation forms quite as marked a feature in the living and breathing description of the *Lady of the Lake*, as the mural picturesqueness of the crags and precipices which the vegetation half-conceals; and this, be it remembered, is not an ordinary characteristic of the Scottish Highlands, though true to nature in the mica-schist region selected by Scott as the scene of his story. After employing, in describing the rocks near Loch Katrine, well-nigh half the vocabulary of the architect,—spires, pyramids, and pinnacles,—towers, turrets, domes, and battlements,—cupolas, minarets, pagodas, and mosques,—he goes on to say,—

> ' Nor were these earth-born castles bare,
> Nor lacked they many a banner fair;
> For from their shivered brows displayed,
> Far o'er the unfathomable glade,
> All twinkling with the dewdrop's sheen,
> The brier-rose fell in streamers green,
> And creeping shrubs of thousand dyes
> Waved in the west wood's summer sighs.
> Boon nature scattered free and wild
> Each plant or flower, the mountain's child.
> Here eglantine embalmed the air,
> Hawthorn and hazel mingled there,
> The primrose pale and violet flower
> Found in each cliff a narrow bower;
> Foxglove and nightshade, side by side,
> Emblems of punishment and pride,
> Grouped their dark hues with every stain
> The weather-beaten crags retain,

> With boughs that quaked with every breath;
> Grey birch and aspen wept beneath;
> Aloft the ash and warrior oak
> Cast anchor in the rifted rock;
> And higher yet the pine-tree hung
> His shattered trunk, and frequent flung,
> Where seemed the cliffs to meet on high,
> His boughs athwart the narrow sky.'

Here is there a description of the characteristic vegetation of our richer mica-schist valleys, not more remarkable for its poetic luxuriance than for its strict truth,—truth so strict and literal, that I question whether even the hyper-critic, who looked for but a typical catalogue, could enumerate more than two forms of vegetation prevalent in such districts which it does not include. The *ferns* grow at once singularly rank and delicate in the shade, amid the bosky recesses of the mica-schist; and every damper recess of the rock we find thickly tapestried over by the *mosses* and the *liverworts*.

Passing southwards along the dark surface of Loch Lomond, skirted for rather more than two-thirds of its length by these hills of mica-schist, which confer on its upper reaches a character of mingled picturesqueness and sublimity, we enter, nearly opposite the pastoral village of Luss, on a band of clay-slate,—the last or most modern of the primary formations. It is of no great breadth,—some three or four miles at most; but it runs diagonally across the entire kingdom, from the western shores of Bute, where it disappears under the outer waters of the Firth of Clyde, to near Stonehaven, where we lose it in the German Ocean. We find it associated with a softer style of scenery than the mica-schist. Lacking the multitudinous contortions, and consequent knobs and protuberances, of the schist, it is less picturesque, though scarce less beautiful; nor is its beauty devoid of an ennobling mixture of the sublime. The gracefully-contoured hills that rise immediately behind Luss, with their recluse withdrawing valley,—the green rolling meadow on which the village is built,—and in front the bolder and finer islands of the lake,—belong all to the clay-slate, and compose a very characteristic landscape. Dunkeld, Comrie, and the fine country to the north and west of Callander, including Loch Vennachar, with many a scene besides of a character intermediate, as becomes their place, between the Highlands and Lowlands, occur in the belt of clay-slate that sweeps in its diagonal course from sea to sea. Leaving Luss behind us, we enter, ere quitting the lakes, on what is unmistakably the low country. The framework of the land before us and on either hand, with that of about one-half the lower islands of Loch Lomond, is all formed of the Old Red Sandstone; and what Byron would perhaps term the 'domestic beauties' of the prospect,—swelling hills ploughed to the top, green lanes, rich meadows, and woods whose rectilinear edges still tell of the planter's line, —bear evidence to the fact. The land, however, is that of Buchanan and of Smollett. Both were born on the Old Red Sandstone here; and the latter, in his well-known description of the lake, in *Humphrey Clinker,*—the product of a time when descriptions of Scottish scenery were less common than they are now,—places in the foreground, in a style unmistakable from their truth, the features of this Lowland formation, which, in his age, was unfurnished with a name. 'I have seen,' he says, 'the Lago di Garda, Albano, De Vico, Bolsena, and Geneva, and, upon my honour, prefer Loch Lomond to them all,—a preference which is certainly owing to the verdant islands that seem to float upon its surface, affording the most enchanting objects of repose to the excursive view. Nor are the banks destitute of beauties which even partake of the sublime. On this side they display a variety of woodland, corn-fields, and pasture, with several agreeable villas emerging, as it were, out of the lake, till, at some distance, the prospect terminates in huge mountains covered with heath. Everything here is romantic beyond imagination : the country is justly termed the Arcadia of Scotland.' In the corn-fields here, the woodlands, and the pastures, we recognise the *Lowland* features of the Old Red placed prominently in the foreground; and in the huge mountains in the distance, the bolder *Highland* features of the clay-slate and the mica-schist. In still journeying southwards, we skirt the banks of the Leven,—the stream which connects the waters of the lake with those of the Clyde, and which, for the greater part of its course, runs over an Old Red Sandstone of the same age as that of Balruddery, Carmylie, and Turin, and which presents as its characteristic organism, the Cephalaspis. And nowhere in Scotland, as is well shown in Smollett's classical Ode, is there a more thoroughly Lowland river.

> 'Pure stream, in whose transparent wave
> My youthful limbs I wont to lave ;
> No torrents stain thy limpid source,
> No rocks impede thy dimpling course,
> That sweetly warbles o'er its bed,
> With white, round, polished pebbles spread.
> Devolving from thy parent lake,
> A charming maze thy waters make,
> By bowers of birch and groves of pine,
> And hedges flower'd with eglantine.'

Ere, however, closing our journey of a day, which introduces us to so interesting an epitome of the scenery of the primary rocks and the Scottish Highlands, we are startled in the midst of the low country by scenery which seems to be that of the Highlands repeated, but on a smaller scale, and, if I may so express myself, in a more *mannered* style. We pass over a narrow belt of the traprocks, which, like the stratified deposits of this part of the kingdom,—clay-slate and Old Red Sandstone,—runs from sea to sea, and which, including in its range the Campsie and the Ochil hills, is here represented by the picturesque double-peaked rock which bears the ancient fortalice of Dumbarton,—the castle which, according to Jeanie Deans's friend Mr. Archibald, was always given in keeping to the best man in Scotland,—at one time to Sir William Wallace, at another to the Duke of Argyll.

The depth of the primary stratified rocks, which in Scotland must be very great, has been variously estimated by geologists,—as low as five and as high as ten miles,—evidence enough, did we require any such, that there must be some degree of obscurity in the data on which the calculations regarding it have been founded. It is always extremely difficult to estimate the thickness of even a clay-slate or quartz-rock deposit in a mountainous country, where the centres of disturbance are numerous and involved; and in gneiss and mica-schist,—always greatly contorted deposits,—the difficulty is so enhanced, that what begins as calculation usually ends as guess. But we at least know that it can be no thin series of deposits, however much their strata may be contorted, or however often repeated, that covers, in highly inclined positions, tracts of country so extended as even those which we find covered by them in the Scotch Highlands. In crossing the four primary stratified deposits,—clay-slate, mica-schist, quartz-rock, and gneiss,—at right angles with the line in which they traverse the country in the southern division of the Highlands, we find them occupying, as from near Crieff to Fort-Augustus, a tract rather more than sixty miles across; and in crossing at the same angle the northern division of the Highlands,—as from Glen Urquhart to the middle reaches of Loch Carron,—we find a tract of nearly forty miles occupied by the gneiss alone. The question is one on which I would not choose to dogmatize; but an estimate that gave to our Scottish primary rocks an aggregate thickness of from six to eight miles I would not regard as by any means too high. A more vexed question, however, and a still more doubtful one, respects their formation. In what form, and under what circumstances, it has been often asked, and very variously answered, were these stratified primary rocks deposited?

They exhibit with almost equal prominence two distinct classes of phenomena,—an igneous class and an aqueous class; and are as intimately associated with the Pleistocene rocks by the one, as with the sedimentary rocks by the other. I have seen in the same quarry of quartz-rock, one set of strata as decidedly chemical in their texture as porphyry or hypersthene, and another intermingling set as decidedly mechanical as grauwacke or conglomerate. I have seen, too, in the same gneiss rock, the minute plates of mica, so abundant in this formation, arranged between the layers as decidedly on the sedimentary principle as in a micaceous sandstone, and in the layers themselves as decidedly on the crystalline principle as in granite. And this compound character of the gneiss may be regarded as the general one, with, of course, certain exceptions in all the primary stratified rocks: the condition of their stratification is mechanical and sedimentary, but the condition of the strata themselves igneous and chemical. How were these variously-blended characters first induced? The geologists of one school tell us that the primary formations originally existed as ordinary sedimentary rocks, but that they have since been altered by the action of intense heat, and that, while the stratification remains as an evidence of their first condition, the texture of the strata indicates the igneous change which has passed over them; while the geologists of another school hold that their first deposition took place under circumstances essentially unlike any which now exist, on at least the surface of our planet, and that their mineralogical conditions were, in consequence, originally different from those of any deposition taking place at the present time, or in any of the later geological ages. I am inclined to hold that there is a wide segment of truth embodied in the views of the metamorphists; but there seems to be also a segment of truth on the other side; and so I must likewise hold with their antagonists, that there existed long periods in the history of the earth in which there obtained conditions of things entirely different from any which obtain now,—periods during which life, either animal or vegetable, could not have existed on our planet; and further, that the sedimentary rocks of this early age may have derived, even in the forming, a constitution and texture which, in present circumstances, sedimentary rocks cannot receive.

The scientific world is subject, like the worlds of politics and trade, to its periods of action and re-action. Those who hold that the earth was once a molten mass throughout,—nay, that at a certain not very profound depth its matter may be still in an incandescent state,—may have perhaps driven their theory too far; and the current at present seems to have set in against them. Mr. Hopkins' profound deductions on the phenomena of Precession and Nutation have been held to establish that the crust of the earth is at present a solid unyielding mass to the depth of at least a thousand miles from the surface. 'Nay, there is nothing in this inquiry,' says Professor Nichol, in referring, in his late admirable work, *The Planetary System*, to the problem of Mr. Hopkins,—'there is nothing in this inquiry rendering it impossible that the globe is solid throughout; and assuredly a distinct negative is given to a whole class of prevalent geological conceptions, on grounds vastly more solid than any which appear to sustain them.' And I find Sir Charles Lyell, in the latest edition of his *Principles*,— that of last year,—suggesting the existence of a circle of superficial action in the earth's crust, quite sufficient to account for an intermittent igneous activity altogether independent of central heat, and which might go on by fits and starts for ever, and be as powerful a million of years hence as in those incalculably ancient times when our Scottish gneiss was in the forming. Accepting the theory of Sir Humphrey Davy, of an unoxidized metallic nucleus of the globe, capable of being oxidized all around its periphery by the percolation of water, and of evolving heat enough in the process to melt the surrounding rocks, he thus provides plutonic, metamorphic, volcanic agencies; and whereas Sir Humphrey Davy held, that when a thick crust of oxide had once formed in this way, it served to shut out the water, and the chemical action became in consequence more and more languid, till it altogether ceased, Sir Charles finds, in another but harmonizing theory, an expedient for re-invigorating the slumbering plutonic forces, and thus, after a period of repose, renewing their activity. The oxygen of the water is, of course, the oxidizing agent; but water also contains hydrogen, and hydrogen is a de-oxidizing agent. 'When the oxidizing process was going on,' says Sir Charles, 'much hydrogen would of necessity be evolved: it would permeate the crust of the earth, and be stored up for ages in fissures and caverns; and whenever it happened to come in contact with the metallic oxides at a high temperature, the reduction of these oxides would be the necessary result.' And we have thus a circle of forces,—oxidization of the metallic basis to evolve the plutonic agencies, and de-oxidization of the oxides to produce the metallic basis again. The process would somewhat resemble that on which the movement of the steam-engine depends, and in which water is first expanded into steam, and then the steam in turn condensed into water, and thus the action of the engine kept up.

Now, I need not here say how thoroughly I respect the judgment and admire the genius of Sir Charles Lyell,—one of the greatest of geologists, and a man of whom Scotland may well be proud; nor need I say how much of pleasure and instruction I owe to the rich and eloquent writings of Professor Nichol. But, like Job's younger friend, I too must take the liberty of showing forth my opinion, and of giving expression to a conviction, on grounds of which my audience must judge, that both Sir Charles and the Professor have suffered the re-action wave to carry them too far.

Mr. Charles M'Laren, in a popular digest of Mr. Hopkins' deductions, which first appeared, if I remember aright, in the *Scotsman* newspaper, and then in Jameson's *Philosophical Journal*, referred, with his characteristic caution, to the narrowness of the base on which they rested. 'Mr. Hopkins' conclusion no doubt rests,' he said, 'on a narrow enough basis. It is somewhat like an estimate of the distance of the stars deduced from a difference of one or two seconds in their apparent position,—a difference scarcely distinguishable from errors of observation.' Let us, however, waive the doubt implied in this remark, however important we may deem it, and grant, for the argument's sake, that the base *is* sufficiently broad for the superstructure erected upon it. Let us freely grant, after first availing ourselves of Mr. M'Laren's protest, and placing it on record, that that equatorial ring, thirteen miles in thickness, which, by disturbing the balance of the earth, is the cause of the phenomena of Precession and Nutation, must be attached to a consolidated crust of at least a thousand miles in thickness, in order to account for the extreme slowness of the peculiar movement which it induces. But let us then inquire how it happens that this equatorial ring at all exists. If our earth was always the stiff, rigid, unyielding mass that it is now,—a huge metallic ball, bearing, like the rusty ball of a cannon, its crust of oxide,—how comes it that its form so entirely belies its history? Its form tells that it also, like the cannon-ball, was once in a viscid state, and that its diurnal motion on its axis, when in this state of viscidity, elongated it, through the operation of a well-known law, at the equator, and flattened it at the poles, and made it altogether the oblate spheroid which all experience demonstrates it to be. It may be urged, however, that this form of our planet, which seems to speak so unequivocally of law, may, after all, be but accident. If so, it must be singular. What say the other planets? Of these, the form of three may be at least approximately, and that of one exactly, ascertained. Venus, Mars, Saturn, are all, like our earth, oblate spheroids, flattened at their poles, and elongated at their equators. Their substance must have been spun out by their rotatory motion in exactly the line in which, as in the earth, that motion is greatest. But while we can only approximately determine the values of the equatorial and polar diameters of these three planets, in one great planet, Jupiter, we can ascertain them scarce less exactly than in our own earth; we can gauge, and measure, and fix the proportions which *his* equatorial ring bears to his general mass. With a diameter about eleven times larger than that of our planet, and rotating on his axis in less than *half* the time, the motion of the surface at his equator must be more than twenty times greater than that of the earth's equatorial surface, and his equatorial ring ought, even in proportion to his huge bulk, to be more than twenty times as massive. And what is the fact? While the thickness of the equatorial ring of the earth is only equal to about one three-hundredth part of the earth's diameter, the equatorial ring of Jupiter is equal to about the one fourteenth or fifteenth part of *his* diameter. It is, as the integrity of the law demands, more than twenty times greater in proportion to his mass than the earth's equatorial ring, and absolutely more than two thousand times greater. Here, then, is demonstration that the oblate sphericity of the earth is a consequence of the earth's diurnal motion on its axis; nor is it possible that it could have received this form when in a solid state. A glass ball made to revolve on a spindle when in a state of viscidity elongates equatorially, and flattens at its poles; but if allowed to cool in its original form as a sphere, it retains its perfect sphericity without change, let us whirl it as rapidly as we may; and no mechanic ever dreams of increasing the disk of a grindstone simply by turning it round. The earth, then, when it assumed its present form, could not have been a solidified mass, like the glass sphere when cooled down, or like the grindstone.

But is it not possible, it may be asked, that the diurnal motion may so act on the depositions taking place in the sea and forming sedimentary rock, or on a region of igneous action interposed between the oxidized crust of the earth and its solid metallic nucleus, and forming plutonic or igneous rock,—is it not possible that, in the course of vastly-extended periods, the earth may have taken its form under the influence of the motion exerted on sedimentary deposition and plutonic intrusion and upheaval? Nay, what, we ask in reply, are the facts? Does the diurnal motion exercise any influence, even the slightest, on deposition or plutonic intrusion? The laws of deposition are few, simple, and well known. The denuding and transporting agencies are floods, tides, waves, icebergs. The sea has its currents, the land its rivers; but while some of these flow from the poles towards the equator, others flow from the equator towards the poles, uninfluenced by the rotatory motion; and the vast depth and extent of the equatorial seas show that the ratio of deposition is not greater in them than in the seas of the temperate regions. We have, indeed, in the arctic and antarctic currents, and the icebergs which they bear, agents of denudation and transport permanent in the present state of things, which bring detrital matter from the higher towards the lower latitudes; but they stop far short of the tropics; they have no connexion with the rotatory motion; and their influence on the form of the earth must be infinitely slight; nay, even were the case otherwise, instead of tending to the formation of an equatorial ring, they would lead to the production of two rings widely distinct from the equator. And, judging from what appears, we must hold that the laws of plutonic intrusion or upheaval, though more obscure than those of deposition, operate quite as independently of the earth's rotatory motion. Were the case otherwise, the mountain systems of the world, and all the great continents, would be clustered at the equator; and the great lands and great oceans of our planet, instead of running, as they do, in so remarkable a manner, from south to north, would range, like the belts of Jupiter, from east to west. There is no escape for us from the inevitable conclusion that our globe received its form as an oblate spheroid at a time when it existed throughout as a viscid mass. Nor is it unworthy of remark, that the same arrangement through which a fluid earth was moulded into this shape under the impulsion of the rotatory motion, also secured that when that earth came to be covered by a fluid sea, placed under the same impulsive influence, it should cling to it equably, like a well-fitted cloak, without falling off to the poles on the one hand, or accumulating in a belt round the equator at the other.

But time fails, and I cannot follow up this subject to its legitimate conclusions. Allow me, therefore, simply to state, that I must continue to hold, with Humboldt and with Hutton, with Playfair and with Hall, that this solid earth was at one time, from the centre to the circumference, a mass of molten matter. Let us remember,—I employ here the words of Humboldt,—that the great chemist Sir Humphrey Davy, to whom we are indebted for the knowledge of the most combustible metallic substances, renounced his bold chemical hypothesis in his last work (*Consolations of Travel*) as 'inadequate and untenable;' and further, that, with the oblate sphericity of the earth and the planets to be accounted for, those who continue to hold what he rejected will be reduced, if they persist, to the unphilosophical necessity of regarding as a consequence of miracle, a peculiarity of shape easily explicable on the principles of known law.

Now, the fact of a molten earth involves a long series of conditions, each different from all the others, and from the conditions of the present time. It involves the existence of a period in the history of our planet when life, animal or vegetable, was not, and of a succeeding period, when life *began* to be. It involves, too, the ripening of the earth from ages in which its surface was a thin, earthquake-shaken crust, subject to continual sinkings, and to fiery outbursts of the plutonic matter, to ages in which it is the very nature of its noblest inhabitant to calculate on its stability as the surest and most certain of all things. It involves, in short, those successive conditions of life in the geologic ages which, in connexion with what is now Scotland, I have, I am afraid, all too inadequately attempted to set before you in my present course. In fine, the primary rocks, when they underlie to a great thickness, as in our own country, the Palæozoic deposits, I regard as the deposits of a period in which the earth's crust had sufficiently cooled down to permit the existence of a sea, with the necessary denuding agencies,—waves and currents,—and, in consequence, of deposition also; but in which the internal heat acted so near the surface, that whatever was deposited came, as a matter of course, to be metamorphosed into semi-plutonic forms, that retained only the stratification. I dare not speak of the scenery of the period. We may imagine, however, a dark atmosphere of steam and vapour, which for age after age conceals the face of the sun, and through which the light of moon or star never penetrates; oceans of thermal water heated in a thousand centres to the boiling point; low half-molten islands, dim through the fog, and scarce more fixed than the waves themselves, that heave and tremble under the impulsions of the igneous agencies;

roaring geysers, that ever and anon throw up their intermittent jets of boiling fluid, vapour, and thick steam, from these tremulous lands; and, in the dim outskirts of the scene, the red gleam of fire, shot forth from yawning cracks and deep chasms, and that bears aloft fragments of molten rock and clouds of ashes. But should we continue to linger amid a scene so featureless and wild, or venture adown some yawning opening into the abyss beneath, where all is fiery and yet dark,—a solitary hell, without suffering or sin,—we would do well to commit ourselves to the guidance of a living poet of true faculty,—Thomas Aird,—and see with his eyes, and describe in his verse :—

> 'The awful walls of shadows round might dusky mountains seem,
> But never holy light hath touched an outline with its gleam ;
> 'Tis but the eye's bewildered sense that fain would rest on form,
> And make night's thick blind presence to created shapes conform.
> No stone is moved on mountain here by creeping creature cross'd,
> No lonely harper comes to harp upon this fiery coast ;
> Here all is solemn idleness ; no music here, no jars,
> Where silence guards the coast ere thrill her everlasting bars ;
> No sun here shines on wanton isles ; but o'er the burning sheet
> A rim of restless halo shakes, which marks the internal heat ;
> As in the days of beauteous earth we see, with dazzled sight,
> The red and setting sun o'erflow with rings of welling light.'

END OF THE LECTURES.

NOTE.

'The only shells I ever detected in the brick-clay of Scotland occurred in a deposit in the neighbourhood of St. Andrews, of apparently the same age as the beds at Portobello.'—*Lecture Second, page* 64.

NOTE.—Some time after this statement was made, Mr. Miller devoted himself to a further investigation of the brick-clay beds in the neighbourhood of Portobello, and discovered several species of shells *in situ*, especially great abundance of *Scrobicularia piperata*, which he has described in a paper on the brick-clays, to be published hereafter. They form a very interesting portion of his Museum, now in the University of Edinburgh. 'But for him,' said an accomplished geologist, in talking with me on the subject, 'we would have known nothing whatever of the brick-clays.'—L. M.

APPENDIX.

GANOID SCALES AND RAYS.

THE scales of the ganoid order consist of three plates,—an inner, an outer, and an intervening one. The outer is composed mainly of enamel, and retains, when entire, however long exposed, much of the original dinginess of hue which it bore in the quarry: the inner is a plane of porcelanic-looking bone: the intermediate plate is finely composed of concentric lines, crossed from the centre to the circumference by finely radiating ones; and when, as mostly happens, this middle plate is exposed, the appearance of a mass of scales through the glass is of great beauty. The rays of our soft-finned fish (Malacopterygii), such as the haddock, seem as if cut through at minute distances, and then re-united, though less firmly than where the bone is entire, with the design, it would seem, of giving to the organs of motion which they compose the necessary flexibility, somewhat on the principle that a carpenter cuts half-through with his saw the piece of moulding which he intends bending along some rounded corner, or forcing into some concave. But in the ancient ganoid fish, in which the rays are bare enamelled bones, and necessarily of great rigidity, the joints appear real, not fictitious. We see them cut across into short lengths, a single fin consisting of many hundred pieces; and the problem lay in conceiving how such a fin was to be wrought,—whether, for instance, each detached length was to have its moving ligament; and if so, how a piece of machinery so very complicated and multifarious was to be set and kept in motion. Here, however, I found the problem very simply resolved. The rays of the ganoid fish, like its scales, consist of three plates,— two plates of enamel, one on each side, and an interior plate of bone. Now the joints,—though so well marked, that in rays imbricated on the sides, as in those of the Cheirolepis, the imbricated markings turn the corners, if one may so speak, just as the carvings on a moulding recounter, as a workman would say, at the corners of a building,—are not real joints after all: they reach but through the inflexible enamel, leaving the central plate of bone undivided. Like the rays of the Malacopterygii, they are formed on the principle of the half-sawn moulding. I observed, too, that the inner plate is in every instance considerably narrower than the plates of enamel which rest upon it. In the lateral edges of every ray which composes the inner portion of the fin there must exist a groove, therefore; and in this groove, it is probable, the connecting membrane at one time lay hid, performing, like an invisible hinge, its work unseen.

RECENT BONE-BED IN THE FORMING.

I ONCE found an interesting illustration of the bone-bed, coupled with at least one of the causes to which it owes its origin, in the upper part of the Moray Firth. I had been spending a night at the herring-fishing, on one of the most famous fishing-banks of the east coast of Scotland,—the bank of Guilliam. It is a long, flat ridge of rock that rises to within ten or twelve fathoms of the surface. On its southern edge there is a submarine valley that sinks to at least twice that depth; and in the course of the night our boat drifted from off the rocky ridge, the haunt of the herrings, to the deepest part of the valley, where scarce a herring is ever found. Our nets had, however, brought fish with them from the fishing-ground, sufficient in quantity to sink them to the bottom of the hollow; and in raising them up,—a work of some little exertion,—we found them bedaubed with patches of a stinking, adhesive mud, that, where partially washed on the surface, seemed literally bristling over with minute fish-bones. The muddy bottom of the valley may be regarded as a sort of submarine burial-ground,—an extensive bone-bed in the forming. 'What,' we asked an intelligent old fisherman, 'brings the fish here to die? Have you observed bones here before?' 'I have observed them often,' he said: 'we catch few herrings here; but in winter and spring, when the cold draws the fish from off the shallows into deep water, we catch a great many haddocks and cod in it, and bring up on our lines large lumps of the foul bottom. In spring, when most of the small fish are sickly and out of season, and too weak to lie near the shore, where the water is rough and cold, they take shelter in the deep here in shoals; and thousands of them, as the bones testify, die in the mud, not because they *come* to die in it, but just because their sickly season is also their dying season.' And such seemed to be the true secret of the accumulation. The fish resorted to this place of shelter, not in order that they might die, but that they might live; just as people go to poor-houses and hospitals with a similar intention, and yet die in them at times notwithstanding. And hence, I doubt not, in most instances those accumulations of fish-bones which men accustomed to the use of the trawl-net find in detached spots of bottom, when in other parts, not less frequented by fish in the milder seasons, not a single bone is to be found, and which have been described as dying places. The dying places,—the deep burial-grounds of the sea's finny inhabitants,—will be found almost always to prove their places of shelter. And hence, it is probable, many of the bone-beds of the geologist.

DIPTERUS MACROLEPIDOTUS ABUNDANT IN THE BANNISKIRK OLD RED OF CAITHNESS.

LET the reader imagine a fish delicately carved in ivory, and then crusted with a smooth shining enamel, not less hard than that which covers the human teeth, but thickly dotted with minute puncturings, as if stippled all over with the point of a fine needle;—let him imagine the enamelled rays lying so thickly in the fins, that no connecting membrane appears, and that each individual ray consists of numerous pseudo-joints, so rounded at their terminations, that each joint seems a small oblong scale, or each ray, rather, a string of oval beads;—in due harmony with the rounded joints, let him imagine the scales of a circular form, and so regularly laid on, that the ruler ranges along them in three different ways,—from head to tail, parallel to the deeply-marked lateral line, and in slant angles across the body;—immediately under the gill-covers, which consist, as in the sturgeon, of but a single plate apiece, let him imagine two strong pectoral fins of an angular form, with an interior angle in each covered with small scales, and the rays, as in the case of the tail, forming but a fringe around it;—let him imagine the ventral fins, which lie far adown the body, of an exactly similar pattern,—angular projections covered with scales in the centre, and fringed on two of their edges with rays;—exactly opposite to these, let there occur an anterior dorsal fin of comparatively small size, and then exactly opposite to the anal fin a posterior dorsal of at least twice the size of the other; let the anal fin be also large and sweeping, extending for a considerable way under the tail, which must, like the tails of all the more ancient fish, be formed mainly on the under side, the vertebral column running on to its termination;—and the fish so formed will be a fair representation of the ancient Dipterus. Presenting externally in its original state no fragment of skin or membrane, and with even its most flexible organs sheathed in enamelled bone, it must have very much resembled a fish carved in ivory. What chiefly struck me in the examination was the peculiar structure of the ventral fins,—the hind paws of the creature, if I may so speak. Their internal angle of scales imparts to them an appearance of very considerable strength,—such an appearance as that presented by the hind fins of the Ichthyosaurus, which, as shown by a lately-discovered specimen, were furnished on the outer edges by a fringe of cartilaginous rays; and I deemed it interesting thus to mark the true fish approximating in structure, ere the reptilia yet existed, to the reptile type. The young frog, when in its transition state, gets its legs fully developed, and yet for some little time thereafter retains its tail. The Dipterus seems to have been a fish formed on this sort of transition plan.

FOSSIL-WOOD OF THE OOLITE AT HELMSDALE, SUTHERLAND.

WHAT first strikes the observer in the appearance of the fossil-wood of this coast is the great distinctness with which the annual layers are marked. The harder lines of tissue, formed in the end of autumn, stand out as distinctly on the weathered surfaces as we see them in pieces of dressed deal that have been exposed for a series of years to the light and the air. The winters of the Oolitic period in this northern locality must have been sufficiently severe to have given a thorough check to vegetation. We are next struck by the great inequality of size in these layers, as we find them shown in separate specimens. I brought with me one specimen in which there is a single layer nearly half an inch in breadth, and another in which, in no greater space, there occur fourteen different layers. The tree to which the one belonged must have been increasing in bulk fourteen times more rapidly than that of the other. Occasionally, too, we find very considerable diversities of size in the layers of the same specimen. One year added to its bulk nearly half an inch; in another it increased scarce an eighth part. Then, as now, there must have been genial seasons, in which there luxuriated a rich-leaved vegetation, and other seasons of a severer cast, in which vegetation languished. My microscope, a botanist's, was of no great power; but, by using its three glasses together, and carefully grinding down small patches of the weathered wood till it began to darken, I could ascertain with certainty, from the structure of the cellular tissue, what, indeed, seemed sufficiently apparent to the naked eye from the general appearance of the specimens, that they all belonged to the coniferæ. When viewed longitudinally, I could discern the elongated cells lying side by side, and the medullary rays stretching at right angles across; but my glass lacked power to show the glandular dots. When viewed transversely, the regularly reticulated texture of the coniferæ was very apparent. A bluish-grey limestone adhered to some of the specimens, and bore evidence in the same track. It abounded in cones and fragments of cones, in what seemed minute needle-shaped leaves, and in thin detached pieces of bark, like those which fall off in scales from the rind of so many of the coniferæ. The limestone bore also its frequent fragments of fern. There seemed nothing lacking to restore the picture. There rose before me a solemn forest of pines, deep, shaggy, and sombre; its opening slopes and withdrawing vistas were cheered by the lighter green of the bracken; and far beyond, where the coast terminated, and the feathery tree-tops were relieved against the dark blue of the sea, a long line of surf tumbled incessantly over a continuous reef of coral.

I picked up one very fine specimen, which, though it weighed nearly a hundredweight, I resolved on getting transported to Edinburgh, and which now lies on the floor before me. It is a transverse cut of a portion of a large tree, including the pith, and measures twenty-three inches across. In the sections of trees figured by Mr. Witham in his interesting and valuable work, the original structure seems much disorganized : a granular radiating spar occupies the greater portion of the interior ; and the tissue is found to exist in but detached portions. Here, on the contrary, the tissue exists unbroken from the pith to the outer ring. We may see one annual circle succeeding another in the average proportion of about ten per inch ; and though we cannot reckon them continuously, for there are darker shades in which they disappear,—shades which the polisher of the marble-cutter may yet succeed in dissipating,—the number of the whole must rather exceed than fall short of a hundred. However obscure the geologist may be in his eras generally, here at least is the record of one century. But how were its years filled? I sat beside the root of a newly-felled fir some six or eight seasons ago, and amused myself, when the severed vessels were throwing up their turpentine in minute transparent globules, in reckoning the years by the rings, from the bark inward. Here, I said, is the year in which the Reform Bill passed ; and this the year in which Canning died ; and this the year of the great commercial crisis ; and this the year of Waterloo ; and this of the burning of Moscow. The yearly rings of the Oolite have no such indices of recollection attached to them : we see their record in the marble, but know no more of contemporary history than that, when forests showed their fringes of lighter green on the hill-sides, and cell and fibre swelled under the rind, the promptings of instinct were busy all around and beneath,—that the pearly ammonite raised its tiny sails to the breeze, as the belemnite, with its many arms, shot past below,—that nameless birds mingled with flying reptiles,—and that, while the fierce crocodile watched in his pool for prey, the gigantic iguanodon stretched his long length of eighty feet in the sand. But who shall reveal the higher history of the time? The reign of war and of death had commenced long before; and who shall assert that moral evil had not long before cast its blighting shadows over the universe,—that there had not been that war in heaven in which the uncreated angel had overthrown the dragon,—or that unhappy intelligences did not wander, 'seeking rest, but finding none,' in an earth of 'waste places,' whose future sovereign still lay hid in the deep purposes of Eternity?

ASTREA OF THE OOLITE, SUTHERLAND.

THE same deposit in which I found the wood embedded contains large masses of coral, all apparently of one species, —not a branching coral, but of the kind which consists of large stone-like masses covered on the surface with stellular impressions, framed in polygons, and which composes the genus Astrea. I picked up one very fine specimen, which I have since got cut through and polished. It presents a polygonal partitioning, of a delicate cream-colour, that somewhat resembles the cells of a honeycomb. Each cell is filled with a brownish ground of carbonate of lime; and on this ground of brown there is a cream-coloured star, composed of rays that proceed from the centre to the sides. One of these corals measured two feet and a half across in one direction, by two feet in another; and if it grew as slowly as some of its order in the present scene of things, its living existence must have stretched over a term of not less extent than that of its contemporary the pine of the hundred rings. Some of the masses seem as if still adhering to the rocks on which they originally grew; the pentagonal cells are still open, as if the inhabitants died but yesterday; and the star-like lines inside still retain their original character of thin partitions, radiating outwards and upwards from a depressed centre. In other instances they have been torn from their places, and lie upturned in the shale, amid broken shells and fragments of wood. I brought with me one curious specimen perforated by an ancient pholas: the cavity exactly resembles those cavities of the existing Lithodomus shell which fretted so many of the calcareous masses that lay scattered on the beach on every side; but it is shut firmly up by the indurated shale in which the specimen itself had lain buried, and a fragment of carbonized wood lies embedded in the entrance. The cave is curtained across by a wall of masonry immensely more ancient than that which converted into a prison the cave of the Seven Sleepers.

RECENT TYPES OF FOSSILS.

AN imagination curious to re-erect and restore finds assistance of no uninteresting kind among the pools and beneath the bunches of sea-weed which we find scattered, at the fall of the tide, over the surface of the Navidale deposits. One very minute pool of sea-water, scarcely thrice the size of a common washing basin, and scarcely half a foot in depth, furnished me with recent types of well-nigh all the fossils that lay embedded for several feet around it; though there were few places in the bed where these lay more thickly. Three beautiful sea anemones,—two of crimson, and the third of a greenish-buff colour,—stretched out their sentient petals along the sides; and the minute currents around them showed that they were all employed in their proper trade of winnowing the water for its animalcular contents, working that they might live. One of the three had fixed its crimson base on the white surface of a fossil coral; the pentagonal cavities, out of each of which a creature of resembling form had once stretched its slim body and still minuter petals, to agitate the water with similar currents, were lying open around it. In another corner of the pool a sea-urchin was slowly dragging himself up the slope, with all his red fleshy hawsers that could be brought to bear, and all his nearer handspokes hard strained in the work. His progress resembled that of the famous Russian boulder, transported for so many miles to make a pedestal for the statue of Peter the Great; with this difference, however, that here it was the boulder itself that was plying the handspokes and tightening the ropes. And lo! from the plane over which he moved there projected the remains of creatures of similar type;—the rock was strewed with fossil handspokes, greater in bulk than his, and somewhat diverse in form, but whose general identity of character it was impossible to mistake. The spines of echini, fretted with lines of projections somewhat in the style of the pinnacles of a Gothic building, lie as thickly in this deposit as in any deposit of the Chalk itself. The pool had its zoophytes of the arborescent form,—the rock its flustra; the pool had its cluster of minute muscles,—the rock its scallops and ostrea; the pool had its buccinidæ,—the rock its numerous whorls of some nameless turreted shell; the pool had its cluster of serpulæ,—the serpulæ lay so thick in the rock, as to compose, in some layers, no inconsiderable proportion of its substance.

BRORA COAL-FIELD OTHER THAN THE TRUE COAL MEASURES.

A COAL-FIELD in other than the true Coal Measures is always an object of peculiar interest to the geologist; and the coal-field of Brora is, in at least one respect, one of the most remarkable of these with which geologists are yet acquainted. The seams of the well-known Bovey coal of South Devon,—a lignite of the Tertiary,—are described as of greater depth; but it burns so imperfectly, and emits so offensive an odour, that, though used by some of the poorer cottagers in the neighbourhood, and some of the local potteries, it never became, nor can become, an article of commerce. It is curious merely as an immense accumulation of vegetable matter passing into the mineral state,—as, shall I venture to say, a sort of half-mineralized *peat* of the Tertiary,—a *peat moss* that, instead of overlying, underlies the diluvium. In the Brora coal, as might be inferred from its much greater age, the process of mineralization is more complete; and it furnishes, if I mistake not, the only instance in which a coal newer than that of the carboniferous era has been wrought for centuries, and made an article of trade. There were pits opened at Brora as early as the year 1598: they were re-opened at various intermediate periods in the seventeenth and eighteenth centuries; on one occasion, in the middle part of the latter, by Williams, the author of a *Natural History of the Mineral Kingdom*, which has been characterized by Lyell as 'a work of great merit for its day;' and during twelve years of the present century, from 1814 to 1826, there were extracted from but a single pit in this field no fewer than seventy thousand tons of coal. The Oolitic coal-field of Sutherland stands out in prominent relief amid the ligneous deposits that derive their origin from the later geological floras. And yet its commercial history does not serve to show that the speculations of the miner may be safely pursued in connexion with any other than that one wonderful flora which has done so much more for man, with its coal and its iron, than all the gold mines of the world. The Brora workings were at no time more than barely remunerative; and the fact that they were often opened to be as often abandoned, shows that they must have occasionally fallen somewhat below even this low line. Latterly, at least, it was rather the deficient quality of the coal that militated against the speculation, than any deficiency in the quantity found. It burned freely, and threw out a powerful flame; but it was accompanied by a peculiar odour, that seemed to tell rather of the vegetable of which it had been originally composed than of the mineral into which it had been converted, and then sunk into a white light ash, which every breath of air sent floating over carpets and furniture. And so, when brought into competition, in our northern ports, with the coal of the Mid-Lothian and English fields, it failed to take the market. The speculation of Williams was singularly unlucky. He became lessee of the entire field about the year 1764, and wrought it for nearly five years. There occurs near the centre of the main seam a band of pyritiferous concretions, which here, as elsewhere, have the quality of taking fire spontaneously when exposed in heaps to air and moisture, and which his miners had not been sufficiently careful in excluding from the coal. A cargo which he had shipped from Portsoy, in Banffshire, took fire in this way, in consequence, it has been said, of the vessel springing a leak; and such was the alarm excited among his customers, that they declined dealing with him any longer for a commodity so dangerous. And so, after an ineffectual struggle, he had to relinquish his lease.

LONDON MUSEUM OF ECONOMIC GEOLOGY.

IN the Museum of Economic Geology now in the course of forming in London, there are specimens exhibited of not only the various rude materials of art furnished by the mine and the quarry, but also of what these can be converted into by the chemist and the mechanic. Not only does it show the gifts of the mineral kingdom to man, but the uses also to which man has applied them. The rough and unpromising block of marble stands side by side with the exquisitely polished and delicately-sculptured vase. The bracelet of glittering steel, scarcely of less value than if wrought in gold, ranges in striking contrast with the earthy, umbry nodule of clay-ironstone. There are series of specimens, too, illustrative of the various changes which an earth or metal assumes in its progress through the workshop or the laboratory. Here, for instance, is the ironstone nodule,—there the roasted ore,—yonder the fused mass; the wrought bar succeeds; then comes the rudely-blocked ornament or implement; and, last of all, the exquisitely finished piece of work, as we find it in the cutler's warehouse or the jeweller's shop.

I am not aware whether the museum also exhibits its sets of specimens illustrative of substances elaborated, not by man, but by nature herself, and elaborated, if one may so speak, on the principle of serial processes and succeeding stages. The arrangement in many cases would have to proceed, no doubt, on a basis of hypothesis; but the cases would also be many in which the hypothesis would at least not seem a forced one. It was suggested to me on the Brora coal-field, that the process through which nature makes coal might be strikingly illustrated in this style. One might almost venture to begin one's serial collection with a well-selected piece of fresh peat, containing its fragments of wood, its few blackened reeds, its fern-stalks, and its club-mosses. Another specimen of more solid and homogeneous structure, and darker hue, cut from the bottom of some deep morass, might be placed second in the series. Then might come a first specimen of Bovey coal, taken from under its eight or ten feet of Tertiary clay,—a specimen of light and friable texture, and that exhibited more of its original vegetable qualities than of its acquired mineral ones. A second specimen, brought from a deeper bed of the same deposit, might be chosen by the darker brown of its colour, and its nearer approximation to the structure of pit-coal. The Oolitic coal of the Brora or Yorkshire field might furnish at least two specimens more. And thus the collector might pass on, by easy gradations, to the true Coal Measures, and down through these to the deeply-seated anthracite of Ireland, or the still more deeply-seated anthracite of America,—not altogether so assured of his arrangement, perhaps, as in dealing with the processes of the laboratory or the workshop, but at least tolerably sure that both chemists and naturalists would find fewer reasons to challenge than to confirm it.

BRORA PEAT-MOSSES OF THE OOLITE.

THE Brora field, so various in its deposits, must have existed in many various states,—now covered by salt water, now by fresh,—now underlying some sluggish estuary,—now presenting, perchance, a superaqueous surface, darkened by accumulations of vegetable matter,—and now, again, let down into the green depths of the sea. To realize such a change as the last, one has but to cross the Moray Firth at this point to the opposite land, and there see a peat-moss covered, during stream tides, by from two to three fathoms of water, and partially overlaid by a stratum of sea-sand, charged with its characteristic shells. It is a small coal-bed, kneaded out and laid by, though still in its state of extremest unripeness,—a coal-bed in the raw material; and there are not a few such on the coasts of both Great Britain and Ireland. Professor Fleming's description of the submerged forests of the Firths of Forth and Tay must be familiar to many of my readers. They must have heard, too, through the far-known *Principles* of Lyell, of the submerged forests of Lancashire. 'In passing over *Black Sod* Bay, in a clear, calm morning,' says a late tourist in Erris and Tyrawly, 'I could see, fathoms down, the roots of trees that seemed of the same sort as are every day dug out of our bogs.' Now, we do not know that the Oolite had properly its peat-mosses. The climate, though its pines had their well-marked annual rings, seems, judging from its other productions, to have been warmer than those in which peat now accumulates; but there can be no doubt that both it and the true Coal Measures must have had *their vast accumulations of vegetable matter formed, in many instances, on the spot on which the vegetable matter grew;* and no one surely need ask a better definition of a peat-moss. A peat-moss, in the present state of things, is simply an accumulation of vegetable matter formed on the spot on which it grew. These, as I have said, we frequently find let down on our coast far beneath the sea-level, and covered up by marine deposits; and the fact furnishes a first and important step in the proposed serial arrangement of coal in the forming. May I not further add, that Professor Johnston of Durham, so well known in the field of geological chemistry, regards all our coal-seams, whether of the Carboniferous period or of the Oolitic, as mere beds of ancient peat, mineralized in the laboratory of nature?

QUARRY OF BRAAMBURY UPPER OOLITE, SUTHERLAND.

ON entering the quarry hollowed on the southern eminence, one is first struck by the character of the broken masses of stone that lie scattered over the excavations. The rubbish abounds in what seem fragments of a very exquisite sculpture. The shells and lignites, which it contains in vast numbers, exist as mere impressions in the white sandstone, and look as if fresh from the chisel of a

Thom or Forrest. But even these masters of their art would confess themselves outdone here in beauty of finish. Their best works don't stand the microscope; whereas the carvings of the Upper Oolite here, though in sandstone, mightily improve under it. The cast of a broken fragment of wood at present before me shows not only the markings of the annual rings, but also the microscopic striæ of the vegetable fibre,—a niceness of impression impossible in any sandstone that had not what the sandstones of this quarry have,—a large mixture of calcareous cement. I remember that, on my first introduction to the excavations of Braambury,—for such is the name of the quarry,—the vast amount of what seemed broken sculpture in the rubbish reminded me of some of Tennant's singularly happy descriptions in his *Dinging down o' the Cathedral*. They seemed memorials of a time when, to the signal detriment of ecclesiastical architecture in Scotland, and all the good solid religion that springs out of sandstone,—

> 'Ilk tirlie-wirlie mament bra,
> That had for centuries ane and a'
> Brankit on bunker or on wa',
> Cam tumblin tap o'er tail . . .
> Whan in ilk kirk the angry folk
> Carv't wark, an arch, an pillar broke.'

I had not a few other recollections of the quarry of Braambury. Nothing can be more interesting to the geologist than its fossils, and nothing more annoying at times to the workman. Occurring often in the wrought stone, they occasion sad gaps and deplorable breaches, where the plane should be smooth or the moulding sharp. I remember laying open on one occasion a beautiful cast that had once been a belemnite, but that had become a mere cavity in which a belemnite might be moulded,—for even this solid fossil, that so doggedly preserves its substance in most other deposits, is absorbed by the sandstone of Braambury. And greatly did I admire its peculiar state of keeping. The smooth cylindrical hollow was partitioned across by two stony diaphragms, thin as bits of drawing-paper; for ere the absorbing process had begun, the fossil had been broken into three pieces by the superincumbent weight, and the minute strips of sand which had filled up the cracks had hardened into stone. The point was sharp and smooth; a rectilinear convex ridge showed the place of the abdominal groove; a cone at the base, lined transversely, represented the chambered shell of the interior. There could not be a more interesting specimen for a museum; but, alas! it occupied the polished plane of a tombstone, just where the *hic jacet* should have been; and though it symbolized the sentence wonderfully well, it was a symbol which I feared few would succeed in interpreting. I pointed it out to a brother workman. 'Ah,' said he, 'you have got one of these terrible tangle-holes; they're the dash'dest things in all the quarry.'

Many a curious thing besides does this quarry contain: boles of trees, that look as if sculptured in the white sandstone, with their gnarled and twisted knots and furrowed rinds; striated reeds of the same brittle material, that seem the fluted columns of architectural models; club-mosses, with their gracefully-disposed branches; rounded stems, scaled like the cones of the fir; impressions of fibrous, sword-shaped leaves, that resemble the leaves of the iris; and the casts of fragmentary masses of timber, deeply fretted by the involved and tortuous gnawings of some marine worm. Such are a few of the sculptured representations of the flora of the period,—things more delicate by a great deal than those carved flowers of Melrose which we find described with such picturesque effect by Sir Walter. And its fauna we see represented quite as interestingly as its flora. Its sculptured Pectens remind us of those of a Grecian frieze; a beautifully-ribbed Cardium has proved a still finer subject for the chisel; its Gryphites stand out in the boldest style of art. One very striking Ammonite (*Ammonite perarmatus*) exhibits a double row of prominent cones, that run along the spiral windings, and give to it the appearance of an Ionic volute ingeniously rusticated; and another Ammonite, that takes its name from the quarry (*Ammonite Braamburiensis*), presents on its smooth, broad surface,—for in form it resembles some of our recent nautili, —the gracefully-involved lines of the internal partitioning, as sharp and distinct as if traced on copper by the burin. The traveller explores and examines, and finds the rude excavation on the hill-side converted into the studio of some wonderful sculptor. In the quarry opened on the other eminence there are similar appearances presented; but the stone is softer, and the impressions less sharp.

GLACIERS AND MORAINES OF SUTHERLAND.

LET us mark the abrupt and imposing character of the hills. They rise dark, lofty, and bare, and show—to employ a graphic Highland phrase—their bones sticking through the skin. They must have been well swept, surely; and as they are composed mainly of Old Red Sandstone conglomerate in this locality,—for we have left behind us the granitic hills of Navidale and Loth,—their sweepings, could we but find them, would have doubtless a well-marked character. And now let us turn to appearances of another kind. We stand on the polished surface of the rock, with its rectilinear grooves and scratches, and, when we look *upwards* along the lines, see the mountains and the valley; but what see we when we look *downwards* along the lines? Something exceedingly curious indeed;—double and triple ranges of miniature hills, composed of boulders and gravel,—the veritable conglomerate sweepings of the mountain-slopes and the valley, mixed with sweepings of the more distant primary hills that rise behind. There they lie, in lines that preserve such a rude parallelism to the steep range from which they were originally scraped, as the waves that rebound from a seaward barrier of cliff maintain to the line of the barrier. Varying from thirty to forty feet in height, and steep and pyramidal, in the cross section, as roofs of houses, they run in continuous undulating lines of from a hundred yards to half a mile in length. Three such lines, with their intervening valleys, occur between the base of Braambury Hill and the village of Brora, like inner, outer, and middle mounds of circumvallation in an ancient hill-fort. If one steadily rakes, with the edge of one's moist palm, the scattered crumbs on a polished tea-table, they form, of course, into irregular lines, presenting in the transverse section a rudely angular form; and in the direction in which they have been swept, the moisture from the palm furrows the mahogany with minute streaks of dimness. The illustration is one on the smallest scale possible. But if the palm be tolerably moist, the crumbs tolerably abundant, and the polish of the mahogany brought brightly out, and if we rake into rude parallelism in this way, line after line from the front of some platter or bread-dish, upturned to represent the line of hills, we shall have provided ourselves with no very inadequate model of the phenomena of Braambury. But what palm of inconceivable weight, breadth, and strength, could have been employed here in thus raking the débris into lines of hills half a mile in length by at least thirty feet in height, and in pressing into smoothness, as it passed, the asperities of the solid rocks below? The reader has already anticipated the reply. We have before us indications of an ancient glacier the most unequivocal that are to be found perhaps anywhere in the kingdom: there is not a condition or accompaniment wanting. I have had my doubts regarding glacial agency in Scotland: but after visiting this locality a twelvemonth ago, I found doubt impossible; and I would now fain recommend the sceptical to suspend their ultimate decision on the point, until such time as they shall have acquainted themselves with the grooved and polished rocks of Braambury, and the parallel moraines that stretch out around its base.

I had lacked time, during my visit of the previous season, to examine the moraines that lie in the opening of the valley higher up, and now set out to explore them. The day had become exceedingly pleasant: a few cottony-looking wreaths of mist still mottled the hills, and the sky overhead was still laden with clouds; but ever and anon the sun broke out in hasty glimpses, that went flashing across the dark moors, now lighting up some bosky recess or abrupt cliff, now casting into strong prominence some insulated moraine. The hollow between Braambury and the hills is occupied, as I have said, by an extensive morass, in which the inhabitants of the neighbourhood dig their winter fuel, and which we find fretted, in consequence, by numerous rectilinear cavities, filled with an inky water, and roughened and darkened on its drier swellings with innumerable parallelograms of peat. I passed an opening in which there were no fewer than five gnarled, short-stemmed fir-trees, laid bare. They lay clustered together, as if uprooted and thrown down by some tremendous hurricane,—presenting exactly such appearances as I have seen in the woods of Cromarty after the hurricane of November 1830, when, in less than an hour, three thousand full-grown trees were blown down in one not very extensive wood, and lay heaped on some of the more exposed eminences in groups of six and eight. A few hundred yards from the prostrate trees there rises, amid the morass, a solitary moraine. I could see its gravelly root extending downwards under the peat which, in the slow course of ages, had accumulated around it, and found the conviction pressing upon me, that many centuries ago, when the five prostrate pines were living denizens of the forest, and the moss which now enveloped them had not yet formed, this insulated hill must have raised its heathy ridge over the trees, and borne the marks of an antiquity apparently not less remote than those which it bears now. And then, long ere the hill itself had formed, the same remark must have applied with at least equal force to the Oolitic rock below. We see that, when overlaid by the ponderous ice, it must have been exactly the same sort of hard, brittle sandstone it is at the present moment. As shown by the slim partitionings that divide internally its casts of Belemnites, it must have hardened ere its fossils were absorbed; and, as shown by its polished and striated surfaces, its fossils must have been absorbed ere the glacier slid over it. We see laid bare in the lines of the striæ, the casts of Gryphites, Pectens, and Terebratulæ; we see further, that the hollows which they formed were weak places in the stone, and that the ice, breaking through, had crushed into them the minute fragments of which their roofs had been composed; and so infer from the appearances, that the newer Oolite of Sutherland must have been as firm a building-stone in the ages of the glaciers as it is now.

As we approach the valley of the Brora, we see a long, well-marked moraine sweeping in a curved line along the base of the hill that forms its northern boundary of entrance, and are again reminded, by the general parallelism of moraine and hill, of the reversed wave thrown back from a barrier of rock. In the gorge of the valley, immediately below where the river expands into a fine wild lake, we find the moraines very abundant, but preserving no regularity of line. They exist as a broken, cockling sea of miniature hills; and, to follow up the twice-used illustration, remind one of rebounding waves at the opening of a rocky bay, where the lines meet and cross, and break one another into fragments. Like many of the other moraines of the Highlands, they were of mark enough to attract the notice of the old imaginative Celtæ, who called them *Tomhans*, and believed them to be haunts of the fairies,—domiciles whose enchanted places of entrance might be discovered on just one night of the year, but which no man, not desirous of becoming a denizen of fairyland, would do well to enter. The lake above is a fine lonely sheet of water, fringed with birch, and overlooked by many a green uninhabited spot, dimly barred by the plough. A range of stern, solemn-looking hills rises steep and precipitous on either hand; while a single picturesque hill, with abrupt sides and a tabular summit, terminates the upward vista some six or eight miles away. I saw in one reedy bay a whole community of water-lilies opening their broad white petals and golden stamens to the light; and, wishing to possess myself of one that grew nearer the shore than any of the others, and having no such companion as Cowper's dog Beau to bring it me, I cut a long switch of birch, and struck sharply at the stem, that I might decapitate it, and then steer it to land. But the blow, though repeated and re-repeated, fell short; and I had drawn my last, when up there started from the bottom a splendid lily, two-thirds developed,—a true Venus, that, rising from the water, looked up to the light, neck-deep, with the rest. The agitation occasioned by the strokes had burst the calyx, and, true to its nature, up the prematurely-liberated flower had sprung. The image which the incident furnished mingled curiously with my attempted restorations of the ancient state of the valley. The delicate lily, rising to the surface in its quiet, sheltered bay, during a bright glimpse of sunshine, formed an interesting point of contrast to what seemed a fast foaming river of ice, that rose on the hillsides more than half their height, and swept downwards, till where it terminated in the plain, in an abrupt moving precipice, that ploughed before it, in its irresistible march, huge hills of gravel and stone.

LEVEL STEPPES OF RUSSIA, AND THEORY OF MORAINES.

IN the level steppes of Russia, where the traveller may journey without seeing a hill for weeks together, the rocks have their grooved and polished surfaces. And even in localities where there *are* hills, the hills not unfrequently merely add to the difficulty. The lofty top of Schehallion, for instance, is grooved and polished; and, pray, from what neighbouring eminence could the glacier have descended on it? Extreme, however, as the difficulties that environ the phenomena may seem, they have been manfully met by Agassiz, and dealt with in a style in which only a man of genius could have dealt with anything. And if difficulties still attend his theory, there are at least other difficulties which it ingeniously obviates; and it seems but right, at all events, to give it generous entertainment and a fair trial, until such time as it may be found untenable, or until at least something better turns up to set in its place.

The flat steppes of Russia have, I have said, their groovings and polishings: they have also their moraine; and so enormous is the extent of the latter, that for week after week the traveller may find it stretching through the central wilds of the empire, on and on without apparent termination, by North Novogorod towards Pinsk, as far as the confines of Silesia. It exists as a broad belt of erratic blocks, mingled with heaps of gravel, and resembles, from its linear continuity, the scattered remains of some such vast wall as that which protected of old the Chinese frontier from the Tartar. And here, says Agassiz, is the moraine of a glacier that had for its centre no group of local eminences, no vanished Alps of the Frozen Ocean, but the North Pole itself. The ice of the Southern Pole advances as far. Could we but reverse the conditions of the two poles, the northern icy barrier would extend to the English Channel, and the whole British islands would lie enveloped in one vast glacial winding-sheet, that, overlying the summits of our hills, would furrow with its parallel striæ even the granitic top of Schehallion.

A complete reversal of the conditions of the two poles would account, doubtless, for many of the phenomena existing in connexion with the boulder-clay, which seem otherwise so inexplicable. But is the reversal itself possible? A Laplace or Lagrange could perhaps answer the question. This much, however, men of lower attainments may know: that the meteorological condition of the two poles are very different,—the icy barrier advancing, in the case of the one, many degrees nearer the equator than it does in the case of the other; that their astronomical condition is also very different, the sun being many millions of miles nearer the one in winter, and nearer the other in summer. It may be known, further, that these astronomical conditions are in a state of gradual change; that, so far at least as human observation extends, the change has been steadily progressing in one direction; that should it but continue, a time must inevitably arrive when their astronomical circumstances shall be wholly reversed,—a time when the sun shall look down upon our northern hemisphere in *aphelion* in winter, and in *perihelion* in summer. True, we do not yet know that the meteorological differences of the poles depend on their astronomical differences, or whether the gradual diminution in the eccentricity of the earth's orbit, which has been lessening these latter differences ever since astronomers registered their observations, may not be like the change in the ecliptic,—the result of a mere oscillation, limited to a few degrees.

Let us, however, conclude the case to be otherwise: let us deem the oscillations in the earth's orbit to be so great as to involve an alternate progress in the sun, between his two foci; let us further infer a dependence between his place in each and the meteorological condition of the poles. We stand, let us suppose, on the summit of a hill; but, as if an immense wedge had been thrust between our feet and the soil, we rise to a higher elevation on an inclined plane of ice, and look over a frozen continent, enlivened by no winding arms of the sea, and bounded by no shore. In the words of Coleridge,—

> 'The ice is here; the ice is there;
> The ice is all around;
> It cracks and growls, and roars and howls,
> A wild and ceaseless sound.'

It is summer; and the sun, in *perihelion*, looks down with intense glare on the rugged surface. There is a ceaseless dash of streams that come leaping from the more exposed ridges, as they shrink and lessen in the heat, or patter from the sunlit pinnacles, like rain from the eaves of a roof in a thunder-shower. They disappear in cracks and fissures; and we may hear the sound, rising from where they break themselves, far beneath, in chill caverns and gloomy recesses, where, even at this season, at noon the temperature rises but little above the freezing-point, and sinks far beneath it every evening as the sun declines. The night shall scarce have come on when all these water-courses shall be bound up by the frost, and the melted accumulations which they precipitated into the fissures beneath shall be converted into expansive wedges of ice, under the influence of which the whole ice-continent shall be moving slowly onwards over the buried land. Millions of millions of wedges shall ply their work during the night on every square mile of surface, and the coming day shall prepare its millions of millions more. There is thus a slow but steady motion induced towards the open space where the huge glacier terminates; the rocks far below grind down into a clayey paste, as the ponderous mass goes crushing over them,—deliberate, when at its quickest, as the hour-hand of a timepiece; and vast fragments are borne away from submerged peaks and precipices by the enclasping solid, just as ordinary streams bear along their fragments of rock and stone from the banks and ridges that lie most exposed to the sweep of their currents. All around, according to Milton,

> 'A frozen continent
> Lies dark and wild, beat by perpetual storms.'

Not a peak of our higher hills appears: all are enveloped in their cerements of cold and death. Even along the flanks of the gigantic Alps, the groovings and polishings rise, says Agassiz, to an elevation of nine thousand feet; and then, and not before, do we find the pinnacles that overlooked the scene standing up sharp and unworn. If we ask a varied prospect, we must remove from our present stand, to where Mont Blanc and his compeers raise their white summits over the line of the horizon, to give earnest of a buried continent, or to where the smoke and fire of Hecla ascends amid the level from a dripping crater of ice.

CROMARTY.

CROMARTY,—my own especial manor, which I have so often beat over, but not yet half exhausted,—presents to the geologist one of the most interesting centres of exploration in Scotland. Does he wish thoroughly to study our Scotch Lias, Upper and Lower, with the Oolitic member which immediately overlies it ?—then let him remove to Cromarty, and study it there. Is he solicitous to acquaint himself with the fossils of the Lower[1] Old Red Sandstone in that state of finest preservation in which the microscope finds most of beauty and finish in them ?—then let him by all

---
[1] Now ascertained to be *Middle*.

means settle at Cromarty. Is he wishful of knowing much about the last elevated of our granite hill-ranges,—a range newer apparently than many of our south-country traps?—let him not hesitate to take lodgings at Cromarty. Is he curious regarding our boulder-clay?—let him set himself carefully to examine the splendid sections which it presents in the neighbourhood of Cromarty. Does he feel aught of interest in our raised beaches?—then let him come and live upon one at Cromarty. Is he desirous of furnishing himself with a key to the geology of the north of Scotland generally? —in no place will he be able to possess himself of so complete a key as among the upturned strata of Cromarty. Had he to grope his way along a course of discovery, he might find the district yielding up its more interesting phenomena but slowly: to know its Lias deposits thoroughly would be a work of months, and to know its Old Red Sandstone, a work of years; but with some intelligent guide to point out to him the localities to which his attention should be directed, and all in them that has been done and observed already, he would find that much might be accomplished in the course of a single week,—especially in the long calm days of July, when the more exposed shores of the district, with all their insulated stacks and ledges, and all their deepsea caves, may be explored by boat.

CAVES OF CROMARTY, OR THE ART OF SEEING OVER THE ART OF THEORIZING.

WE swept downwards through the noble opening of the Cromarty Firth, and landed under the southern Sutor, on a piece of a rocky beach, overhung by a gloomy semicircular range of precipices. The terminal points of the range stand so far out into the sea, as to render inaccessible, save by boat, or at the fall of ebb in stream tides, the piece of crescent-shaped beach within. Each of the two promontories is occupied by a cave in which the sea at flood stands some ten or twelve feet over the gravel bottom, and there are three other caves in the semicircle, into which the tide has not entered since it fell back from the old coast line. The larger and deeper of the three caves in the semicircular inflection is mainly that which we had landed to explore. It runs a hundred and fifty feet into the granitic rock, in the line of a fault that seems first to have opened some eight or ten feet, and then, leaning back, to have closed its sides atop, forming in this way a long angular hollow. It has borne for centuries the name of the *Doocot* (*i.e.*, Dove-cot) *Cave*, and has been from time immemorial a haunt of pigeons. We approach the opening : there is a rank vegetation springing up in front, where the precipice beetles over, and a small stream comes pattering in detached drops like those of a thunder-shower; and we see luxuriating under it, in vast abundance, the hot, bitter, fleshy-leaved scurvy-grass, of which Cook made such large use, in his voyages, as an anti-scorbutic. The floor is damp and mouldy; the green ropy sides, which rise some five-and-twenty feet ere they close, are thickly furrowed by ridges of stalactites, that become purer and whiter as we retire from the light and the vegetative influences, and present in the deeper recesses of the cave the hue of statuary marble. The last vegetable that appears is a minute delicate moss, about half an inch in length, which slants outwards to the light on the prominences of the sides, and overlies myriads of similar sprigs of moss, long since converted into stone, but which, faithful in death to the ruling law of their lives, still point, like the others, to the free air and sunshine. As we step onwards, we exchange the brightness of noon for the mellower light of evening. A few steps further, and evening has deepened into twilight. We still advance; and twilight gives place to a gloom dusky as that of midnight. We grope on, till the rock closes before us; and, turning round, see the blue waves of the firth through the long, dark vista, as if we viewed them through the tube of some immense telescope. We strike a light. The roof and sides are crusted with white stalactites, that depend from the one like icicles from the eaves of a roof in a severe frost, and stand out from the other in pure, semi-transparent ridges, that resemble the folds of a piece of white drapery dropped from the roof; while the floor below has its rough pavement of stalagmite, that stands up, wherever the drops descend, in rounded prominences, like the bases of columns. The marvel has become somewhat old-fashioned since the days when Buchanan described the dropping cave of Slains,—'where the water, as it descends drop by drop, is converted into pyramids of stone,'—as one of the wonders of Scotland, and deemed it necessary to strengthen the credibility of his statement by adding, that he had been 'informed by persons of undoubted veracity that there existed a similar cave among the Pyrenees.' Here, however, is a puzzle to exercise our ingenuity. Some of the minuter stalactites of the roof, after descending perpendicularly, or at least nearly so, for a few inches, turn up again, and form a hook, to which one may suspend one's watch by the ring; while there are others that form a loop, attached to the roof at both ends. Pray, how could the descending drop have returned upwards to form the hook, or what attractive power could have drawn two drops together, to compose the elliptical curve of the loop? The problem is not quite a simple one. It is sufficiently hard at least, as it has to deal with only half-ounces of rock, to inculcate caution on the theorists who profess to deal with whole continents of similar material. Let us examine somewhat narrowly. Dark as the recess is, and though vegetation fails full fifty feet nearer the entrance than where we now stand, the place is not without its inhabitants. We see among the dewy damps of the roof the glistening threads of some minute spider, stretching in lines or depending in loops. And just look here. Along this loop there runs a single drop. Observe how it descends, with but a slight inclination, for about two inches or so, and then turns round for about three quarters of an inch more; observe further, that along this other loop there trickle two drops, one on each side; that, as a consequence of the balance which they form the one against the other, their descent has a much greater sweep; and that, uniting in the centre, they fall together. We have found a solution of our riddle, and received one proof more of the superiority of the simple art of seeing over the ingenious art of theorizing.

But let us proceed to the proper business of the excursion. We have provided ourselves with tools for digging; and, selecting a spot some thirty feet within the cavern, where the bottom seems composed of a damp dark mould, we set ourselves, with spade and pick-axe, to penetrate to the sea-gravel beneath. The soil yields as easily to the tool as a piece of garden-mould; and turning it up to the light in cubical adhesive masses, we find it consisting of an impalpable brown earth, that exactly resembles raw umber. We have fallen on a bed of pure guano, not quite so rich, perhaps, as that which our agriculturists export from the rocky islets of South America at the rate of about fourteen pounds per ton, for it must have been formed originally of vegetable, not animal matter, and we find that it lacks the strong ammoniacal smell of the guano produced by predacious water-birds; but judging from its appearance, and from the high estimate formed of old of the dung of pigeons as a manure, it must be of value enough to deserve removal from the damp unproductive floor of the Doocot. We find the bed which it composes extending downwards from two to three feet, and filling the cavern from side to side. A rock-gravel lies below, hardened into an imperfect breccia by a ferruginous cement; but the rotting moisture exuded from the guano has been unfavourable, apparently, to the preservation of shells, and we find that it contains nothing organic. We again remove to the inner recesses of the cave. Mark, first, that peculiar appearance along the sides. There stands out, at the height of about four feet from the present floor, what seems a rude projecting cornice of rock-gravel, bound together by the stalactitical cement: the projection at one point somewhat exceeds eighteen inches; and we find it bearing short-stemmed stalagmites atop, just like the rugged pavement below. To use a homely but apt illustration, the appearance is that presented by the lower part of a tallow-candle that had been burning exposed to a current of air, with its grease running down in ridges on the sides, and then spreading out on the margin of the meta-socket, when, after raising it out of the candlestick, we see the lower accumulation projecting from it like a cornice. That line of projecting gravel indicates the level at which the floor of the cavern once stood. If we remove the looser parts of the present floor, we shall find its place indicated by just a similar line of projection. The loose sea-gravel could have adhered to the sides only by having formed the part of the floor in contact with them, until the stalagmitical substance had taken effect upon it, by binding it into a mass, and fixing it where it had lain. Let us break into one of the projections. We find it a true breccia, thickly interspersed with such fragments of shells as we may pick up by hundreds in the neighbouring sea-caves, where the incessant beat of the surf on the hard rocks against which it dashes breaks them into rounded fragments. There, for instance, is a massy little bit of the strong smooth buckie (*Fusus Antiquus*), the largest of British univalves; and there a fragment equally massy of the Icelandic Venus, —both of them productions of the oceans, and of such rivers as the Firths of Cromarty and Dornoch. The materials of the projecting cornice are those of a cavern-beach much exposed to the roll of the surf.

S

Let us now see what our several points of circumstantial evidence amount to. First, then, the bottom of the cave must have stood at one time at least four feet over its present level, and at least fourteen feet over the level of the two sea-caves outside; and yet, just as the sea now covers *them*, must the sea at that remote period have covered *it*. The incessant wave must have resounded along these silent walls as it dashed sullenly onwards, and awakened all their echoes with its harsh rattle as it rolled back. The cavern, at that early time, like all the other deep-sea caves of the coast, could have had no crust of stalactites: its sides and roof must have been as dark and bare as the sides and roofs of the caves outside, where the spray washes away every film of calcareous matter ere it has been deposited for half a day. A sudden elevation of the coast took place, and sudden it must have been, for the loose gravel beach, with its finely comminuted shells, was at once raised beyond the influence of the tides; the stalactitical ridges began to form on the walls, and the sea-gravel to consolidate—where these terminated beneath, and the petrifying water oozed through—into the brecciated cornice. But the waves from the lower line had been encroaching inwards, bit by bit, from the cavern's mouth, washing down the floor to their own reduced level, until they had at length scooped it all out, and left but the hardened projections to mark where it had stood. The cave, though now occupied by only the higher tides, had again become, in some sort, a sea-cave, when a second elevation of the land raised it to its present level. The covering of stalactites thickened along its sides; its minute mosses lived, died, and became marble; and, as age succeeded age, the dark recesses in its roof were cheered by the unerring affections of instinct; and brood after brood, reared with assiduous labour to maturity, went forth, some again to return to their hereditary cells, some to take up their abodes with man. I need scarce say, that the rock, or white-backed dove, is the original of our domestic species.

LINE OF CROMARTY SUTOR.

We find that there leaned against one of the precipices of the Southern Sutor, now washed by the spring-tides, a talus of loose débris, such as we see still leaning against the precipices of the old coast line, and that a calcareous spring, dropping upon it from an upper ledge, had, in the course of years, converted its apex into a hard breccia, and cemented it to the rock, while the base below remained incoherent as at first. During this period it must have lain beyond the sweep of the waves. But a change of level took place; the waves came dashing against the loose débris, and swept it away; and all that now remains of the talus is the consolidated apex, projecting about three feet from the rock. Under another precipice of the Cromarty Sutor we find a line of consolidated débris,—which, like the breccia of the apex, must have been the work of a calcareous spring,— running out about fifty feet into the ebb, where it is altogether impossible it could have formed now. The spring must have flowed downwards for these fifty feet ere it reached the sea; for no sooner could it have touched the latter than its waters would have been diffused and lost; and, even could they have avoided such diffusion, the waves must have prevented the loose gravel on which the calcareous matter acted from remaining sufficiently stationary for a single tide. In each of these cases is the value of the evidence enhanced by the circumstances in which it is given. Both the talus and the brecciated line were formed on a basis of granitic rock, so hard that it strikes fire with steel, and which only a general change of level could have let down to the influence of the tide, or elevated over it.

LESSON TO YOUNG GEOLOGISTS FROM CLAY-BED OF THE NORTHERN SUTOR.

THERE is a stiff blue clay much used in Cromarty and the neighbourhood for rendering the bottom of ponds watertight, and the foundations of cellars impervious to the land-springs, and which, save for its greater tenacity, much resembles the blue boulder-clay of our Coal Measures. It is found in the ebb at half-tide, in a bed varying from eighteen inches to three feet in thickness, which overlies the red boulder-clay, and contains minute fragments of shells, too much broken to be distinguished. I had deemed it a sort of re-formation from strata of a greyish-coloured aluminous shale, which occur in the Old Red Sandstone, and are laid bare in the neighbourhood by the sea. The waves dash against them, and then roll back turbid with the lighter particles, to deposit these in the deep still water outside. But in the place at present occupied by the bed the waves could not have deposited them; it is so much exposed to the surf, that the deposit is gradually wearing down under the friction, and it must have been formed, therefore, at a lower level, and when the sea beat against the ancient beaches. We find further proof that such must have been the case in a soft stratum of grey, shaly sandstone, which rises through the bed, and which is thickly perforated by cells of the *Pholas candidus*, containing in abundance the dead shells, but which has been elevated to a too high place to form any longer a fit habitat for the living animals. I had often examined the fragmentary shells of this clayey layer, in the hope of being able to elicit from them somewhat regarding the history of a deposit older than our present coast line, yet newer than our boulder-clay; but I had hitherto found them in every case too comminuted to yield the necessary evidence. I now succeeded, however, in detecting the same deposit under the Northern Sutor, in the same close neighbourhood as on the Cromarty side to the grey aluminous shale of the Old Red Sandstone, to which it seems to have owed its origin, and abounding in organisms marine and terrestrial. All are recent. I found it containing cones of our common Scotch fir, hazel-nuts, fragments of alder and oak, shells of the common mussel much decomposed, and shells, too, of one of the *Gaper* family (*Myæ arenariæ*), still lying in pairs. The blue adhesive clay in which they are embedded can scarce be distinguished from that of the Lower Lias of Eathie; the *sets* of organisms in the two deposits are also the same,—indicating that their deposition must have taken place under similar conditions. The Lias, like the recent clay, has its cones, its bits of wood, and its marine bivalves lying in pairs; and the sole difference that obtains between them is, that while the cones, and wood, and bivalves of the blue clay are all existences of the present time, the cones, and wood, and bivalves of the Lias represent classes of organic beings that have long since passed into extinction. This clay-bed of the Northern Sutor is one of the best places I know for the young geologist taking his first lesson upon. I deemed it of interest chiefly as corroborative of the fact that our raised beaches on the shores of the Cromarty and Moray Firths belong to exactly the present state of things; nay, that for a very considerable period ere their elevation, when the blue bed was forming in comparatively deep water, both sea and land were stored with their existing productions.

GLACIAL APPEARANCES AT NIGG AND LOGIE.

THERE are two several localities in which, after acquainting one's-self with the glacial moraines of Brora, one may examine with advantage the glacial moraines of the neighbourhood of Cromarty. One of these we find in the parish of Logie, not a hundred yards distant from the great coach road ; the other, in the parish of Nigg, on one of the slopes in which the lofty ridge whose south-western termination forms the Northern Sutor sinks at its north-eastern boundary into the plain of Easter Ross. The Logie moraine extends, for full three quarters of a mile, in a line parallel to the mountain range from which its glacier must have descended. There is a furzy level in front, mottled over with groups of cottages ; the moraine,—thickly planted with fir, and amid whose sheltering hollows the gipsies' tent may be seen in the warmer months, and the houseless Free Church congregation at this inclement season,—forms a long undulating ridge, in what a painter would term the middle ground of the landscape ; while on the swelling acclivities behind, over which the icy plane must have once extended, we see woods, and fields, and stately manor-houses, and, high above all, the heathy mountain ridge, where the sky seems resting on the land. I have not seen the rock laid bare in any part of the cultivated tract which intervenes between the moraine and the upland ridge ; but I entertain little doubt that its surface will be found to bear the characteristic groovings and polishings of the glacial period. The moraines of the hill of Nigg, as might be premised from the lower elevation and narrower slopes of the eminence from which their glacier descended, are of small extent compared with the moraine of Logie. There is, however, one of the number, a beautiful grassy Tomhan, fringed at the base with its thickets of dwarf-birch and hazel, that was deemed commanding enough, in some early age, to be selected as the site of a hill-fort, still known to tradition as the Danish camp, and whose double mound of turf we may still see encircling the summit. It must have been a dreary period when the great glacier of Logie, sloping towards the south, and the lesser glacier of the hill of Nigg, sloping towards the north, saw themselves reflected in the separating strait of sea which at this remote period flowed through the flat valley between. The valley is still occupied for half its length by a sandy estuary, known as the Sands of Nigg, which ere the upheaval of the higher beaches, must have existed as a shallow channel, through which the Firth of Cromarty,—then a double-mouthed arm of the sea, with the hill of Nigg as a mountainous island in the midst,—communicated with the Moray Firth beyond.

PHENOMENA EXPLANATORY OF ACCUMULATIONS OF SHELLS.

THERE are scarce any of the appearances with which the geologist is conversant more mysterious than the immense accumulations of shells which he occasionally finds, as in some parts of Sweden, separated from all extraneous matter, as if they had been subjected to some sifting process,—cleaned, as it were, and laid by; and it has long been a question with him how this sifting process has been effected. The theory that the accumulation had been heaped up by great floods, through which substances of the same specific gravity were huddled together, has been the commonly accepted one; but who ever saw a flood, however great, that did not cast down its mud and its clay among its transported shells, or that had not mingled them, in the process of removal, with its lighter gravels or its sand? In the flat estuary of Nigg, I have seen the sifting process effected through a simple but adequate agency. For about two miles from where the estuary opens into the Cromarty Firth, its wild tracts of yielding sand are thickly occupied by the shells that love such localities,—in especial, by the common cockle. Almost every tide, when the animals are in season, furnishes its vast quantities for the markets of the neighbouring towns, and still the supply keeps up ample as at first. Now the tracts of sand which they inhabit, if not properly quicksands, are at least extremely loose, especially when covered by the tide; and though the creatures succeed, so long as they live, in maintaining their proper place in them within a few inches of the surface, no sooner do they die than the shells begin gradually to sink downwards through the unsolid mass, till, reaching, at the depth of about six feet, a firmer stratum, they there accumulate, and form a continuous bed. The work of accumulation has been going on for many centuries; generation after generation has been dying, to undergo this process of burial,—this process of subarenaceous deposition, if I may so speak; and there are places in the estuary in which the shelly stratum has risen to within a foot or two of the surface. It forms a sort of quarry of shells; and when, about thirty years ago, there was a lime-work established in the neighbourhood, many thousand cart-loads were dug out and burned into lime. I had frequent occasion, some five or six years since, to pass through the estuary at seasons when the mere amateur would have perhaps stayed at home. There runs through it a stream of fresh water, that drains the flat fields and scattered lochans of Easter Ross; and on one of my winter journeys, after a sudden thaw, accompanied by heavy rains, I found the stream swollen to the size of a considerable river, and its bed excavated beneath the usual level some three or four feet, with the sectional line of sand and shells through which it had cut standing up over it like a wall. There was first, reckoning downwards, from a foot to eighteen inches of pure sand; and next, from two feet to two feet and a half of dead shells. The sandy tract all around, for many hundred acres in extent, used to be partially covered with water; every furrow of the ripples, and every depression of the surface, borrowed its full from the receding tide, and, from the general flatness, retained it till its return. But on this occasion, the surface-water had found an unwonted drainage, through the upright sectional front, into the newly excavated bed of the stream. It sank through the upper arenaceous layer as through a filtering stone, and then came rushing through the stratum of shells underneath, brown with the sand which it swept from their interstices. Nor could there be a completer sifting process. For yards and roods together the shells were as thoroughly divested of the sandy matrix in which they had lain as if they had been carefully washed in a sieve. I was bold enough to infer from the phenomenon at the time, that the problem of the unmixed accumulations of shells may be, in at least some cases, not so difficult of solution as has been hitherto supposed. One has but to take for granted conditions such as those of the estuary of Nigg,—the incoherent bed, half a quicksand, and the subarenaceous deposition,—to account for their original production, and the superadded conditions of the surface-water and the free drainage, to account for their after clearance of extraneous matter.

CAUTION TO GEOLOGISTS ON THE FINDING OF REMAINS.

IN consolidated slopes it is not unusual to find remains, animal and vegetable, of no very remote antiquity. I have seen a human skull dug out of the reclining base of a clay bank, once a precipice, fully six feet from under the surface. It might have been deemed, not without a degree of plausibility, the skull of some long-lived contemporary of Enoch, —perchance that of one of the accursed race,—

'Who sinned and died before the avenging flood.'

Nay, a fine theory was in the act of being formed regarding it, which affected the whole deposit ; but, alas! the labourer dug a little further, and struck his pickaxe against an old

Gothic rybat, that lay deeper still. There could be no mistaking the character of the chamfered edge that still bore the marks of the tool, nor that of the square perforation for the lock-bolt; and the rising theory straightway stumbled against it and fell. Both rybat and skull had come from an ancient burying-ground, situated on a projecting angle of the table, and above.

REMARKS ON UNDERLYING CLAY ON LEVEL MOORS.

On level moors, where the rain-water stagnates in pools, and a thin layer of mossy soil produces a scanty covering of heath, we find the underlying clay streaked and spotted with patches of white. As in the spots and streaks of the Red Sandstone formations, Old and New, the colouring matter has been discharged without any accompanying change having taken place in the mechanical structure of the substance which it pervaded; for we find the same mixture of arenaceous and aluminous particles in the white as in the red portions. And the stagnant water above, acidulated, perhaps, by its various vegetable solutions, seems to have been in some way connected with these appearances. In almost every case in which a crack through the clay gives access to the oozing moisture, we find the sides bleached, for several feet downwards, to nearly the colour of pipe-clay; we find the surface, too, when divested of the soil, presenting for yards together the appearance of sheets of half-bleached linen. Now, the peculiar chemistry through which these changes are effected might be found to throw much light on similar phenomena in the older formations. There are quarries in the New Red Sandstone in which almost every mass of stone presents a different shade of colour from that of its neighbouring mass, and quarries in the Old Red, whose strata we find streaked and spotted like pieces of calico. And their variegated aspect seems to have been communicated in every instance, not during deposition, nor after they had been hardened into stone, but when, like the boulder-clay, they had existed in an intermediate state.

TRAVELLED BOULDERS NOT ASSOCIATED WITH CLAY.

ALL the travelled boulders of the north do not seem associated with the clay: we find them occurring, in some instances, in an overlying gravel, and in some instances resting at high levels on the bare rock. I have seen, on the hill of *Fyrish*,—a lofty eminence of the Lower Old Red which overlooks the upper part of the Cromarty Firth,—a boulder of an exceedingly beautiful, sparkling hornblende, reposing on a stratum of yellow sandstone, fully a thousand feet over the sea, where there is not a particle of the clay in sight. We find these travellers furnishing specimens of almost all the primary rocks of the country,—its gneisses, schistose and granitic, its granites, red, white, and grey, its hornblendic and micaceous schists, and occasionally, though more rarely, its traps. The stone most abundant among them, and which is found occurring in the largest masses, is a well-marked granitic gneiss, in which the quartz is white, and the feldspar of a pink colour, and in which the mica, intensely black, exists in oblong accumulations, ranged along the line of stratification in interrupted layers. No rock of the same kind is to be found *in situ* nearer than thirty miles. We find granitic boulders of vast size abundant in the neighbourhood of Tain, especially where the coach-road passes towards the west through a piece of barren moor, and on the range of sea-beach below. One enormous block, of a form somewhat approaching the cubical, is large enough, and seems solid enough, to admit of being hewn into the pedestal of some colossal statue; but instead of being thus appropriated to form *part* of a monument, it has lately been converted of itself into a *whole* monument. When I last passed the way, I found it dedicated, in an inscription of nine-inch letters, '*to the memory of the immortal Scott.*' Nature had dedicated it to the memory of one of her great revolutions ages before; but since the dedicator had determined on adding, in Highland fashion, a stone to the cairn of Sir Walter, it would certainly have been no easy matter to have added to it a nobler one.

GRANITIC GNEISS AND SANDSTONE, WITH THE CONDITIONS OF THEIR UPHEAVAL.

ON entering on the granitic rock, we find the strata, strangely disturbed and contorted, lying, in the course of a few yards, in almost every angle, and dipping in almost every direction. And not only must there have been a complexity of character in the disturbing forces, but the rock on which they acted must have been singularly susceptible of being disturbed. The strata of the sandstone were, at the period of their upheaval, the same brittle, rigid plates of solid stone that they are now. The strata of the granitic gneiss were characterized, on the contrary, during their earlier periods of disturbance, by a yielding flexibility: they were capable of being bent into sharp angles without breaking. We see them running in zig-zag lines along the precipices, now striking downwards, now ascending upwards, now curved like a relaxed Indian bow in one direction, now curved in a contrary one, like the same bow when fully bent. The strata of the sandstone, like a pile of glass-panes laid parallel, existed in a state in which they could be either raised in any given angle, or, if the acting forces were violent and partial, broken up and shivered; whereas the granitic strata existed in the state of the same glass-panes brought to a bright red heat, and capable, from their extreme flexibility, of being bent and twisted in any direction. We find, too, that there occur occasional patches in which the lines of the stratification have been altogether obliterated. We can trace the strata with much distinctness on every side of these; but there is a gradual obscuration of the lines, and we see what was a granitic gneiss in one square yard of rock existing as a compact homogeneous mass in the next. The effect is exactly that which would be produced in the heated panes of my illustration, were the heat kept up until portions of them began to run; and the circumstance serves to throw light on some of the other phenomena of the gneiss. The stone, in its average specimens, is a ternary, consisting of red feldspar, white quartz, and a dingy-coloured mica; but no one, notwithstanding, could mistake it for a true granite. It has its granite *veins*, however; and these *veins*, truly such in some cases, are, in not a few others, mere strata of the gneiss, which have evidently been formed into granite where they lie. There are no marks of injection,—no accompanying disturbance: all their conditions, with the exception of their being true granites, are exactly those of the layers which repose over and under them. Now the homogeneous patches serve, as I have said, to throw light on the secret of the formation of these. In one important respect the granitic rocks differ widely among themselves. Some of them contain potass and soda in such large proportions, and have such a tendency to disintegrate, in consequence, that they furnish much less durable materials for building than the better sandstones; while others, of an almost indestructible quality, are devoid of these salts altogether. Potass and soda form powerful fluxes; and it seems at least natural to infer that, should wide tracts of granitic rock be exposed to an intense but equable heat, the portions of the mass in which the fluxes exist in large proportions must pass into a much higher state of fluidity than the portions in which they are less abundant, or which are altogether devoid of them. Single strata and detached masses might thus come to be in the state of extremest fusion of which their substance was capable, and all their particles, disengaged, might be entering freely into the combinations peculiar to the plutonic rocks, when all around them continued to bear the semi-chemical, semi-mechanical characteristics of the metamorphic ones. Hence it is possibly the origin of some of those granite veins, open above, and terminating below in wedge-like points, which have so puzzled the Huttonians of a former age, and which have been so triumphantly referred to by their opponents as evidences that the granite had been precipitated by some aqueous solution.

SEPTARIA OR CEMENT-STONES OF THE LIAS.

OBSERVE these nodular masses of pale blue limestone, that seem as if they had cracked in some drying process, and had afterwards the cracks carefully filled up with a light-coloured cement. The flaws are occupied by a rich calcareous spar; and in the centre of each mass we find, in most instances, a large ill-preserved Ammonite, which has also its spar-filled cracks and fissures, as if it, too, had been burst asunder by the process which had rent the surrounding matrix. These nodular masses are the characteristic *septaria* or cement-stones of the Lias, so much used in England for making a hard, enduring mortar, that has the quality of setting under water. Their bluish-coloured portions are so largely charged with the argillaceous matter of the bed in which they occur, and contain, besides, so considerable a mixture of iron, that, refusing to slake like common lime, they have to be crushed, after calcination, by mechanical means; while the fossil in the centre, and the semi-transparent spar of the cracks, are composed of matter purely calcareous. And from this peculiar mixture this cement seems to derive those setting qualities which render it of such value.

AMMONITES OF THE NORTHERN LIAS.

THE Ammonites of the upper beds of the Lias approach more to the type of the *Ammonite communis*, being comparatively flat when viewed sectionally, and having the whorls broadly visible, as in the Ionic volute; while the Ammonites of the lower beds approach in type to the *Ammonite heterophyllus*,—each succeeding whorl covering so largely the whorl immediately under it, that the spiral line seems restricted to a minute hollow in the centre, scarce equal in extent, in some specimens, to the twentieth part of the entire area. In other words, the Ammonites of the Upper Lias in this deposit represent, as a group, the true ammonite type; while in the Lower Lias they approach more nearly as a group to the type of the nautilus. And not only are they massier in form, but also absolutely larger in size. I have found Ammonites in the more ponderous septaria, that fully doubled in bulk any I ever saw in the upper shales. We occasionally find nodules that, having formed in the outer rings of these larger shells, somewhat resemble the rims of wheels,—in some cases, wheels of not very diminutive size.

BELEMNITES OF THE NORTHERN LIAS.

WE find the Belemnites of the lower deposit, like its Ammonites, of a bulkier form than those of the upper beds.

The *Belemnites abbreviatus* and *elongatus*, both large, massy species, especially the former, are of common occurrence; while those most abundant in the upper beds are the *Belemnites longissimus* and *penicillatus*, both exceedingly slim species. It is worthy of remark, that Sir R. Murchison, in his list of fossils peculiar to the Lias as developed in the midland counties of England, specifies the *Belemnites penicillatus* as characteristic of its upper, and the *Belemnites abbreviatus* and *elongatus* of its lower division.

CUTTLE-FISH.

Is the reader acquainted with at once the largest and most curious of British Mollusca,—the cuttle-fish,—a creature which stands confessedly at the head of the great natural division to which it belongs? Independently of its intrinsic interest to the naturalist, it bears for the commentator and the man of letters an interest of an extrinsic and reflected kind. No other mollusc occupies so prominent a place in our literature. It is furnished with an ink-bag, from which, when pursued by an enemy, it ejects a dingy carbonaceous fluid, that darkens the water for yards around, and then escapes in the cloud,—like some Homeric hero worsted by his antagonist, but favoured by the gods, or some body of military retreating unseen from a lost field, under the cover of a smoking shot. And there has scarce arisen a controversy since the days of Cicero, in which the cuttle-fish, with its ink-bag, has not furnished some one of the controversialists with an illustration. It has attained to some celebrity, too, on another and altogether different account. That enormous monster, the kraken of Norway, of which our earlier geographers tell such surprising stories, was held to belong to this curious family. And though the monster has disappeared from the treatises of our naturalists for a full half-century, and from the pages of even our more credulous voyagers for at least a century more, it maintained its place as a real existence long enough to be assigned a permanent niche in our literature. It has been described as raising its vast arms out of the water to the height of tall forest-trees, and as stretching its knobbed and warted bulk, roughened with shells, and darkened with seaweed, for roods and furlongs together,—resembling nothing less extensive than some range of rocky skerries on some dangerous coast, or some long chain of sand-banks forming the bar of some great river. It was introduced to the reading world with much circumstantiality of detail, by an old Norwegian bishop (Eric Pontoppidan), as 'an animal the largest in creation, whose body rises above the surface of the water like a mountain, and its arms like the masts of ships.' And one of the French continuators of Buffon,—Denys Montfort,—regarding it as at least a possible existence, has given, in his history of Mollusca, a print of a colossal cuttle-fish hanging at the gunwale of a ship, and twisting its immense arms about the masts and rigging,—a feat which the cuttle-fish of the Indian seas is said sometimes to accomplish, if not with a ship, at least with a canoe. But nowhere does the kraken of Norway look half so imposing or half so poetical as in Milton. In palpable reference to the old bishop's 'largest animal in creation,' we find the poet describing, in one of his finest similes,—

> ' that sea-beast,
> Leviathan, which God of all his works
> Created hugest that swim th' ocean stream:
> Him haply slumb'ring on the Norway foam,
> The pilot of some small night-founder'd skiff,
> Deeming some island, oft, as seamen tell,
> With fixed anchor in his scaly rind,
> Moors by his side under the lee, while night
> Invests the sea, and wished morn delays.'

The existing cuttle-fish of our seas, though vastly less

imposing in its proportions than the kraken of Norway, is, as I have said, a very curious animal,—constituting, as it does, that highest link among Mollusca, in which creatures without a true back-bone or a true brain approach nearest, in completeness of structure and the sagacity of their instincts, to the vertebrata. All my readers on the sea-coast, especially such of them as live near sandy bays, or in the neighbourhood of salmon-fishings, must have frequently seen the species most abundant in our seas,—the common loligo or *strollach* (*Loligo vulgaris*); and almost all of them must have the recollection of having regarded it, when they first stumbled upon it in some solitary walk, as an extraordinary monster, worthy of the first place in a museum. 'The cuttle-fish,' says Kirby, in his Bridgewater Treatise, 'is one of the most wonderful works of the Creator.' We have no creature at all approaching it in size, that departs so widely from the familiar every-day type of animal life, whether developed on the land or in the water.

A man buried to the neck in a sack, and prepared for such a race as Tennant describes in his *Anster Fair*, is an exceedingly strange-looking animal, but not half so strange-looking as a *strollach*. Let us just try to improve him into one, and give, in this way, some idea of the animal to those unacquainted with it. First, then, the sack must be brought to a point at the bottom, as if the legs were sewed up tightly together, and the corners left projecting so as to form two flobby fins; and further, the sack must be a sack of pink, thickly speckled with red, and tolerably open at the other end, where the neck and head protrude. So much for the changes on the sack; but the changes on the parts that rise out of the sack must be of a much more extraordinary character. We must first obliterate the face, and then, fixing on the crown of the head a large beak of black horn, crooked as that of the parrot, we must remove the mouth to the opening between the mandibles. Around the broad base of the beak must we insert a circular ring of brain, as if this part of the animal had no other vocation than to take care of the mouth and its pertinents; and around the circular brain must we plant, as if on the coronal ring of the head, no fewer than ten long arms, each furnished with double rows of concave suckers, that resemble cups arranged on the plane of a narrow table. The *tout ensemble* must serve to remind one of the head of some Indian chief bearing a crown of tall feathers; and directly below the crown, where the cheeks, or rather the ears, had been, we must fix two immense eyes, huge enough to occupy what had been the whole sides of the face. Though the brain of an ordinary-sized loligo be scarcely larger than a ring for the little finger, its eyes are scarce smaller than those of an ox. To complete our cuttle-fish, we must insist as a condition that, when in motion, the metamorphosed sack-racer must either walk head downwards on his arms, or glide, like a boy descending an inclined plane on ice, feet foremost, with the point of his sack first, and his beak and arms last; or, in other words, that, reversing every ordinary circumstance of voluntary motion, he must make a snout or cut-water of his feet, and a long trailing tail of his arms and head. The cuttle-fish, when walking, always walks with its mouth nearer the earth than any other part of either head or body, and when swimming, always follows its tail, instead of being followed by it.

This last curious condition, though doubtless, on the whole, the best adapted to the conformation and instincts of the creature, often proves fatal to it, especially in calm weather and quiet inland firths, when not a ripple breaks upon the shore, to warn that the shore is near. An enemy appears; the creature ejects its cloud of ink, like a sharp-shooter discharging his rifle ere he retreats; and then, darting away tail foremost under the cover, it grounds itself high upon the beach, and perishes there. Few men have walked much along the shores of a sheltered bay without witnessing a catastrophe of this kind. The last loligo I saw strand itself in this way was a large and very vigorous animal. The day was extremely calm; I heard a peculiar sound,—a *squelch*, if I may employ such a word; and *there*, a few yards away, was a loligo nearly two feet in length, high and dry upon the pebbles. I laid hold of it by the sheath or sack; and the loligo, in turn, laid hold of the pebbles, just as I have seen a boy, when borne off against his will by a stronger than himself, grasping fast to projecting door-posts and furniture. The pebbles were hard, smooth, and heavy, but the creature raised them with ease, by twining its flexile arms around them, and then forming a vacuum in each of its suckers. I subjected one of my hands to its grasp, and it seized fast hold; but though the suckers were still employed, it employed them on a different principle. Around the circular rim of each there is a fringe of minute thorns, hooked somewhat like those of the wild rose. In fastening on the hard smooth pebbles, these were overtopped by a fleshy membrane, much in the manner that the cushions of a cat's paw overtop its claws, when the animal is in a state of tranquillity; and, by means of the projecting membrane, the hollow inside was rendered air-tight, and the vacuum completed; but in dealing with the hand, a soft substance, the thorns were laid bare, like the claws of the cat when stretched out in anger, and at least a thousand minute prickles were fixed in the skin at once. They failed to penetrate it, for they were short, and individually not strong, but acting together and by hundreds, they took at least a very firm hold.

What follows the reader may deem barbarous; but the men who gulp down at a sitting half a hundred live oysters to gratify their taste, will surely forgive me the destruction of a single mollusc to gratify my curiosity. I cut open the sack of the creature with a sharp penknife, and laid bare the viscera. What a sight for Harvey when prosecuting, in the earlier stages, his grand discovery of the circulation! *There,* in the centre, was the *yellow* muscular heart propelling into the transparent tubular arteries the *yellow* blood. Beat—beat—beat; I could see the whole as in a glass model; and all I lacked were powers of vision nice enough to enable me to detect the fluid passing through the minuter arterial branches, and then returning by the veins to the two other hearts of the creature; for, strange to say, it is furnished with three. There is the yellow heart in the centre, and lying altogether detached from it, two other darker-coloured hearts at the sides! I cut a little deeper. *There* was the gizzard-like stomach, filled with fragments of minute mussel and crab shells; and *there*, inserted in the spongy, conical, yellowish-coloured liver, and somewhat resembling in form a Florence flask, the ink-bag distended, with its deep dark *sepia*,—the identical pigment sold under that name in our colour-shops, and so extensively used in landscape drawing by the limner. I once saw a pool of water, within the chamber of a salmon-wear, darkened by this substance almost to the consistence of ink. Where the bottom was laid dry, some fifteen or twenty cuttle-fish lay dead, some of them green, some blue, some yellow; for it is one of the characteristics of the creature that, in passing into a state of decomposition, it assumes a succession of brilliant colours; but at one of the sides of the chamber, where there was a shallow pool, six or eight individuals, the sole survivors of the shoal, still retained their original pink tint, freckled with red, and went darting about in panic terror within their narrow confines, emitting ink at almost every dart, until the whole pool had become a deep solution of *sepia*. But I digress.

I next laid open the huge eyes of the stranded cuttle-fish. They were curious organs,—more simple in their structure than those of any quadruped, or even any fish, with which I am acquainted, but well adapted, I doubt not, for the purpose of seeing. A camera-obscura may be described as consisting of two parts,—a lens in front, and a darkened chamber behind; but in both the brute and human eye we find a third part added : there is a lens in the middle, a darkened chamber behind, and a lighted chamber, or rather vestibule, in front. Now this lighted vestibule—the cornea—is wanting in the eye of the cuttle-fish. The lens is placed in front, and the darkened chamber behind; the construction of the organ is that of a common camera-obscura, without aught additional. I found something worthy of remark, too, in the peculiar style in which the chamber is darkened. In the higher animals it may be described as a chamber hung with black velvet; the *pigmentum nigrum* which covers it is of deepest black : but in the cuttle-fish it is a chamber hung with velvet, not of black, but of a dark purple hue; the *pigmentum nigrum* is of a purplish-red colour. There is something curious in marking this, as it were first, departure from an invariable condition of eyes of the more perfect structure, and in them tracing the peculiarity downwards through almost every shade of colour, to the emerald-like eye-specks of the pecten, and the still more rudimental *red* eye-specks of the star-fish. After examining the eyes, I next laid open, in all its length, from the neck to the point of the sack, the dorsal bone of the creature,—its internal shell I should rather say, for bone it has none. The form of the shell in this species is that of a feather equally developed in the web on both sides. It gives rigidity to the body, and furnishes the muscles with a fulcrum; and we find it composed, like all other shells, of a mixture of animal matter and carbonate of lime. In some of the genera it is much more complicated and rigid than in that to which the *strollach* belongs, consisting, instead of one, of numerous plates, and in form somewhat resembling a flat shallop with its cargo rising over the gunwale, or one of the valves of a pearl mussel occupied by the animal. Is my description of this curious creature too lengthy? The young geologist who sets himself to study the fossils of the Oolitic and Cretaceous systems would be all the better for knowing a great deal more regarding it than I have told him here. He will discover that at least one-half the molluscous remains of these deposits, their belemnites, ammonites, nautili, nummulites, baculites, hamites, lituites, turrilites, and scaphites, belonged to the great natural class—singularly rich in its extinct orders and genera, though comparatively poor in its existing ones—which we find represented by the cuttle-fish.

CONGENERS OF THE CUTTLE-FISH, BELEMNITES, ETC.

AMONG its many extinct congeners, the order of the Belemnites was one of not the least curious. It has been remarked, that in the cuttle-fish, as we now find it, a greater number of distinct portions of the organization of creatures belonging to widely-separated divisions of the animal kingdom are to be seen united than in any other animal. Cut off its head immediately below the arms, and we have in the dissevered portion, with its ring of nerve, its central mouth, and its suckers, the true analogue of a star-fish. The radiated zoophyte lies before us. Some of its genera have their plated and jointed antennæ placed above and below the eyes. The creature, so far as these organs give it a character, is no longer a zoophyte, but an insect or crustacean. But then *there* is the soft sac, with its fin-like appendages, the internal shell, and the yellow transparent blood. These are unequivocal characteristics of the mollusc. Yes; but then *there* is a horny beak, and there a muscular gizzard. It must have laid the *bird* under contribution for these. *There* is, besides, a true tongue, and an organ for hearing; and, though one of the chambers be wanting, a singularly large and efficient eye. These organs are all borrowed from the vertebrata. And—as if to secure its claim to originality, not only in its combinations, but in its parts—*there* are its three hearts, and its well-stored ink-bag, chattels that it could scarce have borrowed anywhere. It occupies, according to Cuvier, a sort of central place in the animal kingdom, where roads from all the various divisions converge, and the three hearts and the ink-bag mark, as it were, the point at which they meet. Extensive and wonderful, however, as its combination of parts may seem, its extinct congener the Belemnite added to the number at least one part more. Like that curious gelatinous zoophyte, the Dutch man-of-war (*Physalia*), it was furnished with a sailing apparatus. Not only could it swim tail foremost, and walk head downwards, like our existing cuttle-fish; but it could also raise itself to the surface of the water, and there, spreading out its sail of thin membrane, speed gaily away before the wind. Several of the existing congeners of the creature, such as the *Argonauta Argo*, are sailors still; but, unlike the Belemnite, or its analogue the cuttle-fish, they are furnished with external shells. They are sailors each in its own little boat, whereas the Belemnite was a sailor without a boat,—such a sailor as Franklin was, when, laying himself at full length in the water, he laid hold of the string of an elevated kite during a smart breeze, and, without effort on his own part, was drawn across a small lake by the impulsion of the wind above.

I have full in my view where I write, a shelf occupied with ranges of our Scotch Belemnites of the Lias placed on end, and leaning against the wall, like muskets in an armoury. A second shelf exhibits ranges of our Scotch Belemnites of the Oolite. Ere adverting, however, to their specific differences,—differences which their mode of arrangement renders apparent at a glance, let me select for description an average specimen, as a type of the order. Here, then, is the *Belemnite elongatus*, from the Upper Lias of Eathie. The architect gives the proportions of his columns by a scale of diameters. The height of the Tuscan column is equal to seven, that of the Doric to eight, that of the Ionic to nine, and that of the Corinthian to ten diameters. In describing the proportions of the Belemnite, I shall borrow a hint from the architect, by making my scale one of diameters also; fixing my callipers, not at the base of the shaft, but one-fourth of its entire length higher up. Let the reader imagine a small cylindrical column of brown polished stone, diminishing from the base upwards for three-fourths of its height, much in the same proportions as one of the Grecian columns diminishes, and then in the remaining fourth suddenly sweeping to a point. Its length—eight inches in the present instance—is equal, like that of a Corinthian shaft, to ten of its diameters. Within this solid column we find an internal cone rising from the common base, the whole of which it occupies, and terminating in the apex, at about one-third the height of the whole. It is different in colour and structure from the brown pointed shaft at which it is included. The shaft or column shows as if it had been formed, like a dipped candle, by repeated accessions to its outer surface; whereas the internal cone shows that it has been formed by accessions to its base. The shaft seems to have grown as a tree grows, and exhibits its internal concentric rings crossed by lines radiating from the centre, just as the yearly rings of the tree are crossed by the medullary rays: the internal cone, on the contrary, was reared course after course, as a pyramid is built of ashlar,—with this difference, however, that it was the terminal course of the apex that was laid first, and that every succeeding course was added to the base. The entire Belemnite was originally of greater length than the specimen before us indicates; for the cone extended very considerably beyond the base of the column, and beyond the cone there was a still further prolongation of a kind of horny sheath, composed of the internal shell of an extinct order of cuttle-fish, its substitute for a vertebrate column; just as the existing loligo has its thin elastic pen, and the existing *sepia* its stiffer and more complex bundle of calcareous plates. There are English specimens, in which the characteristic ink-bag may still be found resting on the base of the internal cone, giving evidence at once of the class of animals to which the fossil belonged, and that the column and cone must have been internal, not external, shells. Nature, though liberal to all her creatures, is no spendthrift. We find that to her naked Cephalopoda, such as the *strollach* and the *sepia*, she gives in the ink-bag an ability of hiding themselves in sudden darkness; but that to the shelled creatures of their class, such as the nautilus, she gives no ink-bag. For them the protecting shell is sufficient. The ink-bag of the Belemnite at once shows that it was a cuttle-fish, and that it was naked. Here, in a specimen from the Whitby Lias, we may see the bag still charged with its ink; and so slight is the change induced by untold centuries, in the nature of the carbonaceous substance which composed the latter, that, after having scraped it down, and diluted it with water, we may still use it as a pigment. We find it stated by Buckland, that the tinting of a drawing made with fossil ink at his request by his friend Francis Chantrey was pronounced by a celebrated painter, unacquainted with the secret of its origin, as peculiarly agreeable and well-toned.

But the Belemnite, with its horny prolongation, was not merely a sort of stiffener introduced into the body of the creature to give it rigidity,—as the seamstress introduces, for a similar purpose, bits of wire and whalebone into her pieces of dress, or as the *pen* exists in the *strollach:* the stony column, and its internal cone, constituted, besides, the sailing organs of the creature,—the cone forming its floating apparatus, and the column its ballast. The cone, as I have said, consists of a number of layers, ranged parallel to its base, like courses of ashlar in a pyramid. We find each layer, when detached, exactly resembling a thick patent watch-glass, concave on its under, convex on its upper, surface. Now, each of these formed, in its original state, not a solid mass, but a hollow, thinly partitioned chamber or storey; and, perforating the entire range of storeys from apex to base, there was a cylindrical pipe, just as the reader must have seen the cylindrical case of a turnpike stair passing upwards through the storeys of some ancient tower from bottom to top. And this pipe was the siphuncle or pump through which the creature regulated its specific gravity, and sank to the bottom or rose to the surface, just as it willed. Mr. J. S. Miller, well known for his labours among the Crinoidea, mentions, in his paper on Belemnites, an interesting experiment with regard to the cone. He extracted it carefully from one of his specimens, and then inserting in the hollow of the stony column which it had occupied, a cone of oiled paper filled with cotton, he placed the specimen in water, and found the buoyancy of the cone compensating so completely for the density of the column, that the whole floated. Now, to demonstrate the use of the ballasting column, let us imagine a sail raised over the cone, and the whole sent to sea in a high wind. Has the reader ever sailed, when a boy, his mimic ship, and does he remember how imperative it was that there should be lead on the keel? The stony column is the lead here; and from the form of the creature, as indicated in the entirer specimens, some such internal ballasting seems to have been as essential to preserve its upright position as the lead is to the boy's ship. There are, however, but few of our naturalists who believe, with Mr. J. S. Miller, that the column was originally the dense and solid body it is now. Lamarck held that, like the bone of the existing *sepia*, it was of 'a spongy and cellular texture;' Parkinson, that it was 'porous or cork-like;' and Buckland, that 'the idea of its having been heavy, solid, and stony, while it formed part of a living and floating *sepia*, is contrary to all analogy.' With an eye to the question, I have succeeded in collecting a number of specimens, which, when in their recent state, had been crushed or broken; and I am disposed to hold, from the appearance of the fractures in every case, that, notwithstanding the authorities arrayed against him, Miller's view is the right one. The stony column, though it must have been somewhat less brittle in its recent than in its fossil state,—for it contained its numerous thin plates of horn, tenacious, as is natural to the substance, in a considerable degree,—was yet brittle enough to break across at very low angles, and to exhibit on the side to which the force had been applied, its yawning cracks and fissures, though on the opposite side the wrinkled surface generally indicates a tag of adhesion. In the cases, too, in which the Belemnite had been broken into fragments, I have found every detached portion presenting its hard, sharp angles, and existing as a brittle calcareous body, however soft and chalky the condition of the more delicate shells of the deposit in which it occurred. Nor do I know that analogy is very directly opposed to the supposition that the column might have existed in the creature in its stony state. If two solid calcareous substances, quite as hard and dense as any fossil Belemnite, exist within the head of the recent cod and haddock, why might not one solid calcareous substance have existed within the body of an extinct order of cuttle-fish?

I have found considerable difficulty in classing, according to their species, the Belemnites of the Lias. I soon exhausted the species enumerated as peculiar to the formation by Miller, and found a great many others. They divide naturally into two well-marked families,—the specimens of a numerous family, that, like the *Belemnite elongatus*, are broadest at the base, and diminish as they approach the apex,—while the specimens of a family considerably less numerous, like the *Belemnite fusiformis*, resemble spear-heads, in being broadest near the middle, and in diminishing toward both ends. In subdividing these great families, various principles of classification have been adopted. There are grooves, single in some species, double, and even triple, in others; extending from the apex downwards in some, extending from the base upwards in others; and these have been regarded by Phillips,—the geologist who has most thoroughly studied the subject,—as constituting valuable characteristics not only of species, but of genera and formations. Miller took into account, as principles of classification, not only the general form, but even the comparative transparency or opacity, of the column,—marks selected in accordance with the belief that the column was originally the solid substance it is now. The order furnishes, doubtless, its various marks of specific arrangement. I have even found the hint borrowed from the architect, of taking the proportions of species by their diameters, not without its value. In measuring, for instance, four well-preserved specimens of the *Belemnite abbreviatus*, one of the bulkiest which occurs in our Scotch Lias, and whose average length is six inches, I found that two of the four contained $5\frac{1}{2}$ diameters, one $5\frac{1}{4}$ diameters, and one $5\frac{3}{4}$ diameters; while another bulky Belemnite of the Scotch Oolite, not yet named apparently, whose average length is $3\frac{1}{2}$ inches, contains only $3\frac{1}{2}$ diameters, and strikes at once as specifically different from the others. Equally striking is the specific difference of the *Belemnite elongatus*, which contains from nine to ten diameters,—of another nameless species which contains from twelve to thirteen diameters,—of another which contains from fifteen to sixteen diameters,—of another, agreeing in its proportions with the *Belemnite longissimus* of Miller, which contains from eighteen to twenty diameters,—of another which contains from twenty-three to twenty-four diameters,—and of yet another, long and slender as a heckle-pin, which contains from thirty to thirty-two diameters. My rule of classification must of course be regarded as merely a subsidiary one. There are species which it does not distinguish: it does not distinguish, for instance, the *Belemnite sulcatus* of our Scotch Lias, whose average length is six inches, from the *Belemnite elongatus*, whose average length is eight. Both agree in containing from nine to ten diameters, though in form and appearance they are strikingly different,—the *adjuncatus* being much more pointed at the apex than the other, much more finely polished on the surface, and furnished with a deeper groove. As a subsidiary rule, however, I have found the rule of the diameters a useful one. It has enabled me to form a numerous and discordant assemblage of specimens into distinct groups, the specific identity of which, when thus collected, is at once verified by the eye.

But the reader, unless very thoroughly a geological one, must be of opinion that I have said quite enough about the Belemnite. I may, however, venture to add further, that its place in the geological scale is not without its interest. The periods of the more ancient formations, from the older Silurian to the older New Red Sandstone inclusive, had all passed away ere the order was called into existence. It then sprang into being nearly contemporaneously with the bird and the reptile; and, after existing by myriads during the Oolitic and Cretaceous periods, passed into extinction when the ocean of the Chalk had ceased to exist, and just as quadrupeds of the higher order were on the eve of appearing on the stage, but had not yet appeared. Since the period in which it lived, though geologically modern, the surface of the earth must have witnessed many strange revolutions. There have been Belemnites dug out of the sides of the Himalaya mountains, seventeen thousand feet above the level of the sea.

COPROLITES OF THE LIAS.

LARGE coprolites of peculiar appearance, some of them charged with fish-scales of the ganoid order, are tolerably abundant; and they belonged, I have little doubt, to saurians. When bringing home with me, many years since, a well-marked specimen, I overtook by the way an acquaintance who had passed a considerable part of his life in Dutch Guiana. The thought did not at first occur to me of submitting to him my specimen. As we walked on together, he thrust his hand into his pocket to bring out his handkerchief, and brought out instead a large mass of damaged snuff. 'Ah,' he exclaimed, 'that roguish boy! I was standing with my neighbour the shopkeeper this morning, when he was opening up a cask of snuff that had got spoiled with sea-water; and his boy, seeing my pocket provokingly open I suppose, must have dropped in this huge lump! The joke seems a small one,' he continued, 'but it must be at least rather a natural one. The only other trick of the kind ever played me was by a South American Indian, on the banks of the Demerara: he dropped, unseen, into the pocket of my light nankeen jacket, a piece of sun-baked alligator's dung.' 'What sort of a looking substance was it?' I asked, uncovering my specimen, and submitting it to his examination; 'was it at all like that?' 'Not at all unlike,' was the reply; 'it bore an exactly similar pale yellow tint, as if, like the dung of our sea-birds that swallow and digest fish-bones, it contained abundance of lime; and it was sprinkled over, in the same way, with the glittering enamelled scales of that curious fish the bony pike, so common, as you are aware, in our South American rivers.'

INTRUSIVE DIKES OF EATHIE.

THERE are appearances in connexion with the Lias of Eathie which seem well suited to puzzle the geologist, and which have, in fact, already puzzled geologists not a little. We find them traversed by intrusive dikes of what seems a greyish-coloured trap, extremely obstinate in yielding to the hammer, and which stand up among the softer shales like the walls of some ruined village. They are trap-dikes in every essential except one;—they occur in every possible angle of disagreement with the line of the strata : in some places they enclose the shale in slim insulated strips, as a river encloses its islands : in others they traverse it with minute veins connected with the larger masses, in the way in which granite is so often seen traversing gneiss : in yet others the limestone in contact with them seems positively altered;—the blue nodule has at the line of junction its strip of crystalline white, and the shale assumes an indurated and venous character: the dikes are, in short, trap-dikes in every essential except one; but the wanting essential is of importance enough to constitute the problem in the case;—they are not composed of trap. Some of our mineralogists have been a good deal puzzled by finding *crystals* of sandstone as regular in their planes and angles as if formed of any of the earths, or salts, or metals, whose law it is to build themselves up into little erections correctly mathematical in every point and line; and they have read the mystery by supposing that these sandstone crystals are mere casts moulded in the cavities in which crystals had once existed. The puzzle of the Lias dikes is of an exactly similar kind : they are composed, not of an igneous rock, but of a hard calcareous sandstone, undistinguishable in hand-specimens from an indurated sandstone of the Lower

Oolite, which may be found on the shore beneath Dunrobin, alternating with shale-beds of the period of the Oxford clay. I succeeded in finding in it, on one occasion, a shell in the same state of keeping in which shells are so often found in the resembling rocks of Sutherland, but the species unluckily could not be distinguished. A common microscope at once detects the mechanical character of the mass; and I have learned that Dr. Fleming, after reducing a portion of it, sent him as an igneous rock, to its original sand, simply by submerging it in acid, expressed some little fear lest the sender should not have been quite 'up to trap.'

The explanation of the phenomenon seems rather difficult. There are instances in which what had once been trap-dikes are found existing as mere empty fissures; and other instances in which empty fissures have been filled up by aqueous deposition from above. An instance of the one kind is adduced, as the reader may perhaps remember, in the *Elements* of Lyell, from M'Culloch's *Western Islands;* two contiguous dikes traversing sandstone in Skye are found existing to a considerable depth as mere hollow fissures. An instance of the other kind may be found, says M'Culloch, in a trap rock in Mull, which is traversed by a dike that, among its other miscellaneous contents, encloses the trunk of a tree, converted into brown lignite. In cases of the first kind, the original dike, composed of a substance less suited to resist the action of the weather than the containing rock, has mouldered away, and left the vent from which it issued a mere hollow mould, in which the semblance of a dike might be *cast*, just as the decay and disappearance of the real crystal is supposed to have furnished a mould for the formation of the sandstone one. In cases of the second kind, we see the fictitious dike actually existing: it is the sandstone crystal moulded and consolidated, and, in short, ready for the museum. And we have but to suppose the conditions of the two classes of dikes united,—we have but to suppose that the hollow filled by the aqueous deposition had been previously filled by an igneous injection, —in order to account for all the phenomena of an igneous dike accompanying a merely aqueous one. We can scarce account in this way, however, for the formation of the dikes at Eathie, seeing that the shale in which they are included is of so soft and decaying a character, that no igneous rock could of possibility be more so; nor, even were the case otherwise, could the upper portion of the dikes have existed as open chasms during the period in which the process of decay would have been taking place in the depths below. They would have infallibly filled up with the fragments detached from the sides and edges.

Mr. Strickland, in a paper on the subject in the *Transactions of the London Geological Society*, states the problem very strongly. 'The substance of these dikes is such,' he says, 'that it is impossible to refer them to a purely igneous origin;' and yet, however much 'it may resemble an aqueous product,' it is as impossible to doubt that the dikes themselves are genuine 'intrusive dikes penetrating the Lias shale in all directions.' He adds further, as his ultimate conclusion in the matter, that the 'sedimentary structure of the rock forbids us to refer it to igneous injection from below;' and that, 'notwithstanding the complete resemblance of these intrusive masses to ordinary plutonic dikes, we have no resource left but to refer them to aqueous deposition, filling up fissures which had been previously formed in the Lias.' There is a peculiar rock in the neighbourhood, which throws, I am of opinion, very considerable light on their origin. It is what may be termed a syenitic gneiss, abounding in minute crystals of hornblende, that impart to it a greenish hue; and in one place we find it upheaved so directly among the Lias beds, that it breaks their continuity. It raised them so high on its back, that the denuding agencies laid the back bare by sweeping them away. Let us but imagine that this disturbing rock began to rise under the earlier impulsions of the elevating agencies, and during the deposition of some one of the later secondary formations, as the precursor of the granitic range,—that the superincumbent Lias, already existing in its present consolidated state, opened into yawning rents and fissures over it, as the earth opened in Calabria during the great earthquake,—and that the loose sand and calcareous matter which formed the sea-bottom at the time, borne downwards by the rushing water, suddenly filled up these rents, ere the yielding matrix had time to lose any of its steepness of side or sharpness of edge, which it could not have failed to have done had the process been a slow one. The sandstone dikes, apparently Oolitic, mark, it is probable, the first operations of those upheaving agencies to which we owe the elevation of the granitic wall, and which, ere they accomplished their work, may have been active during occasional intervals for a series of ages. I am not of opinion that the accompanying marks of alteration among the shales and limestones of the beds are sufficiently unequivocal to render imperative some more fiery theory.

CONTEMPORARY AND EXTINCT TYPES OF THE LIFE OF THE TEREBRATULA.

WE find among the earliest bivalves of the Silurian system the delicate Terebratula, with its punctured umbone; we follow it downwards through all the various formations, and see it appearing on each succeeding stage, specifically new, but generally old, until, quitting the rocks with their dead remains, we pass to the existing testacea of our seas, and find among them the ancient Terebratula still extant as a living shell. Contemporary as a genus with every extinct form of animal life, we find it contemporary with the last of created beings also,—contemporary with ourselves; and the Terebratula is but one existence of a class to which, though their generic antiquity may be rather less remote, nearly the same remark applies. The ostrea still exists,—its congener and contemporary the gryphæa has perished; the nautilus survives,—its congener and contemporary the ammonite is long since dead; the cuttle-fish abounds on our shores,—its congener and contemporary the belemnite is to be found in only our rocks. And thus the list runs on. We can scarce glance over a group of fossils, whatever its age, which we do not find divisible into two classes of types,—the types which still remain, and the types which have disappeared. But why the one set of forms should have been so repeatedly called into being, and why the other set should have been suffered to become obsolete, we cannot so much as surmise. Why, it may be asked, should the nautilus continue to exist, and yet the ammonite have ceased with the ocean that deposited the Chalk? or why should we have cuttle-fish in such abundance, and yet no belemnites? or why should not the gryphæa have been reproduced in every succeeding period with the oyster? In visiting some old family library, that has received no accessions to its catalogue for perhaps more than a century, one is interested in marking its more vivacious classes of works,—its Spectators, and Robinson Crusoes, and Shakespeares, and Pilgrim's Progresses, in their first, or at least earlier editions, ranged side by side with obsolete, long-forgotten volumes, their contemporaries, that died on their first appearance, and with whose unfamiliar titles one cannot connect a single association. But it is always easy to say why, in the race of editions, the one class should have been arrested at the very starting-post, and why the other should have gone down to be contemporary with every after production of authorship, until the cultivation of letters shall have ceased. It is otherwise, however, with the geologist. He finds he has exactly the same sort of fact to deal with,

—an immense multiplication of editions, in the case of some particular type of fish, or plant, or shell, and in the case of other types, no after instances of republication; but he finds himself wholly unable to lay hold of any critical canon through which to determine why the one class of types should have been so often republished, or the other so peremptorily suppressed. And yet, were all the circumstances known, it is possible that some such canon might be found to exist. Geology is still in its infancy. Shall a day ever arrive when, in a state of full maturity, it will be able to appeal to its fixed canons, and to say why one certain type of existence was fitted for but one definite stage in the progress of things, and some other certain type fitted, by a peculiar catholicity of adaptation, for every succeeding period?

SIR DAVID BREWSTER ON THE CUTTLE-FISH AND BELEMNITE.

THE following discovery of Sir David Brewster, regarding a marked peculiarity of structure in the eye of the cuttle-fish, now first made public, will be deemed of great interest by all who have learned to admire that inconceivable variety of design in the works of the Infinite Mind which grows upon the inquirer the more he examines, and which, if man were not immortal, it would be an error of his very nature to have the strong existing desire to examine :—

ST. LEONARD'S COLLEGE, ST. ANDREWS.

MY DEAR SIR,—I have been reading, with great pleasure, your interesting account of the cuttle-fish, and was glad to find that you had noticed the singular structure of its eye. During the last twenty years I have dissected literally *hundreds* of cuttle-fish eyes, but I never published my observations on them, in consequence of having found singular discrepancies in the eyes of different species, and having been always expecting from America the eyes of the remarkable varieties which occur there, and which have been repeatedly promised me by American naturalists.

As you will take a great interest in the subject, I shall endeavour to give you some idea of what I have done.

Independent of the peculiarity which you have noticed, of there being no aqueous chamber between the *cornea* and the *lens*, there is no *iris* and no *pupil*, the quantity of light admitted being regulated by the *eyelids*.

The lens itself is of a most singular description. It consists of *two lenses* sticking together, and capable of being separated without injuring either. This structure is unique.

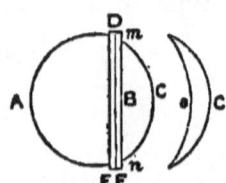

The lens D A E C consists of two, D A E, and a meniscus, *m* C *n*, which is kept close to D A C by a double cartilaginous ring, D E. The dimensions are D E = 0·51 inch, A C = 0·433 inch, A B = 0·3433 inch, B C = 0·09 inch ; *m n* = 0·333 inch. The outer diameter of the front ring, D F, is = 0·59 inch, and its inner diameter = 0·31 inch.

In some indurated lenses I find the lens C to be *doubly convex*, and the surface of the lines D E A, on which it rests, concave. This must have been the lens of a different species.

The fibrous structure of the lens is very remarkable. The laminæ, or coats, of the lens are parallel to D A E and *m* C *n*; and the fibres of the lens D A E diverge from A as a pole, like the meridians of a globe ; and they all terminate, not in another pole, but *in the surface* D E, or that which corresponds with *m o n*. This termination of the whole component fibres of the lens D A E in a surface is quite *unique*, and the mode of converting this rough plane (like a shaving-brush cut across), into a smooth surface, is singularly beautiful. Each elementary coat, or lamina, being composed of fibres, has at its termination in the periphery

APPENDIX.

D E a sort of *selvage*, where all the fibres end; and these selvages, being circles, fill up, as it were, or compose the flat surface of the lens.

The coats, or laminæ, consist of fibres different from those of all other animals. When other lenses harden, they form a solid body, transparent like a *gum;* but the cuttle-fish retains its laminated structure, and shines with all the brilliancy of a *pearl*.

In the *Sæpia Electona* the front lens A separates from B in the line m a b c n, a peculiarity which I have never found in the *Sæpia Loligo*. The diameter A B is larger than m n.

It would be curious to find the lenses in a fossil state.

I have found some lenses of the *Sæpia Loligo* of a paraboloidal form. It is probable that the form of the lens varies with the age of the animal.

When the lenses become indurated, they often exhibit the most beautiful internal reflections, and I have often thought of having them set as brooches. The *pearly* structure is produced by long exposure under ground; and it is almost impossible to distinguish such lenses from *pearls* when the convex part only is shown.—I am, my dear Sir, ever most truly yours,

D. BREWSTER.

To HUGH MILLER, Esq.

THEORY OF THE OCEAN'S LEVEL,

AS AFFECTED BY THE RISING OR SINKING OF THE LAND.

THE mean level of the sea cannot be regarded as a fixed line, unless, during the geologic changes of the past, it has invariably maintained the same distance from the earth's centre. If the earth, in consequence of the expansive influence of a vastly higher temperature than that which in the present era it possesses, was once greatly bulkier than it is now, the line, in proportion to the bulk, would be further removed than it is now from the centre. The sea would stand greatly higher than at its present line. And who that has surveyed the contortions, the bends, the inflections, the ever-recurring rises and falls, of the more ancient stratified rocks, such as our Scotch grauwacke for instance, —bends and inflections that forcibly remind the geologist of the foldings of a loose robe, grown greatly too large for the shrunken body which it covers,—or that has weighed the yet further evidence furnished by the carboniferous vegetation, extra-tropical in character even in Greenland,— who, I say, that has considered this evidence will venture to decide that the earth's temperature was not higher, nor the earth's radius greater, in the days of the Silurian period, or of the Coal Measures, than it is now? And, of course, if the earth's radius was greater, the level line of the sea must have stood higher,—vastly higher, it seems not impossible, than the line now touched by the summits of our highest mountains. Had there been a graduated pole of adamant, equal in length to the radius of the globe, placed in that ocean of the Silurian period in which our Scotch graptolites lived,—a pole with its lower end fixed immoveably at the earth's centre, and its upper end level with the medium surface of the sea,—where, I marvel, would that upper end be now? High, I suspect, in the clouds; nay, in an attenuated atmosphere, to which cloud never now ascends. The graduated markings of the pole, indicatory not merely of how the *tide*, but also of how the *land*, *has fallen*, would, I doubt not, be found more conveniently summable in leagues than in fathoms.

But even setting aside all this as fanciful and extravagant,—even taking it as a given fact (what, I suspect, is no fact at all) that the earth's bulk has not very materially altered, the line of the sea-level may have, notwithstanding, been considerably affected simply by the rise of the land. It is estimated that about one-fourth part of the surface of the globe is occupied, according to the present distribution of oceans and continents, by land, and the remaining three-fourths by water; or, more correctly, that the land is as 1, and the water as 2.76. Let us suppose this fourth part of land *annihilated* to the mean depth of the ocean. Of course, the effect would be, that the ocean, having then to cover four parts, instead of three, would sink, all over the globe, exactly one-fourth part of its mean depth. If the mean depth of the ocean be, as has been estimated, four miles, the fall in its level that would take place, in consequence of this annihilation of the land, would be just a single mile. And, of course, a *creation* of land at the bottom of the sea, which would rise to its surface, would, on the same principle, and in exactly the same ratio, have the effect of *elevating* the ocean level. It would do on a large scale what the pebbles dropped by the crow in the fable into the pitcher did on a small one. Nor must it be forgotten, that though *creation* and *annihilation* are terms which may seem suggestive of the fanciful and the extravagant, there are phenomena exceedingly common in nature which, for all the purposes of my argument, would have exactly the effect of the things which these terms signify. In intense cold, the mercury in a thermometer is confined to the bulb of the instrument; plunged into boiling water, it straightway rises 212 degrees in the tube; and, when a second time subjected to the intense cold, it sinks again into the bulb as at first. So far as mere bulk is concerned, there takes place what is analogous to a creation and annihilation of the quantity of mercury in the tube. Again, if a rod of lead a mile in length be raised in temperature from the freezing point to the point at which water boils, it lengthens rather more than five yards;—what is equal to a creation of five yards of lead-rod has been effected. Cooled down again, however, the five yards are annihilated. A rod of flint-glass of the same length, raised to the same temperature, would stretch out only four feet two inches and rather more than seven lines. All the metals—even platinum—expand more than glass; but were there some deep-lying stratum, five miles in thickness, of that portion of the earth's crust on which Great Britain rests, to be heated 212 degrees above its present temperature, it would at even this comparatively low rate of expansion elevate the island more than twenty feet higher than now over the existing sea-level,—a height fully equal to that of by far the best marked of our ancient coast lines. And if this increase in temperature took place, not in a stratum of the earth's crust *five* miles in thickness, underlying Great Britain, but in a stratum *twenty* miles in thickness, underlying one-fourth the area of the bed of the ocean, the effect would of course be of a reverse character. This *creation* of land at the bottom of the sea would raise the ocean level nearly twenty feet all over the globe, and send the waves dashing around our own shores, against the ancient coast line, as of old.

Nor do I see that the bearing of these consequences on the sea-line,—consequences that would render its level dependent on the elevation or submergence of every continent that has existed, or shall yet exist,—can be set aside, save on the supposition that for every tract of land that rises, another tract of the same area and cubic contents sinks; or, to state the case in other words, and more definitely, that for every protuberance formed within the sea, there is a corresponding hollow formed *also within it* elsewhere. Now, even were it to be granted that for every protuberance which rises on the earth's crust there is a corresponding depression of the surface, which takes place somewhere else (though on what principle this should be granted is not in the least obvious), I do not at all perceive why that depression should always take place *within the sea*. It may take place not on any of the three parts of the earth's surface covered by water, but on that fourth part occupied by land. It may take place on the table-land of a continent. Or, *vice versa*, a hollow formed in the sea, considerable enough to lower the sea's level, may find its counterbalancing protuberance in the further elevation of the interior of some vast tract, such as Asia or New Holland, already raised over the ocean. The submerged continent of the Pacific, which now exists but as a wilderness of scattered atolls, may have been contemporary with that of South America, existing at the time as a flat tract, which simply occupied a certain *area* in the sea; and the hollow which the submergence of the Polynesian land occasioned may possibly have been balanced by the rise of those enormous table-lands of Mexico and the adjacent countries that give to the entire continent in which they are included a mean elevation of more than a thousand feet: or the submergence of that *Atlantis* which was drained by the

APPENDIX.

great rivers of the Wealden period may have been balanced, in like manner, by the rise of the still more extensive table-land of Asia: and in both cases the level of the sea could not fail to be very sensibly lowered. It would have in each instance the area of the submerged continent to occupy; and there would be no corresponding elevation *within its bed*, to balance against the waste by the space which it filled. But why, I repeat, the balancing theory at all? If elevations or depressions can, as has been shown, be mere results of changes of temperature in portions of the earth's crust, why deem it more necessary to hold that there is a refrigerating process taking place under one area, in the exact proportion in which there is a heating process taking place under another, than to hold that when the mercury is rising in the tube of a thermometer, it is sinking in some other tube attached to the instrument, but not visible? The argument, however, is one of those which can be reasoned out more conclusively by lines than by words. It will be found, too, that the lines make out not only a more conclusive, but also a stronger case.

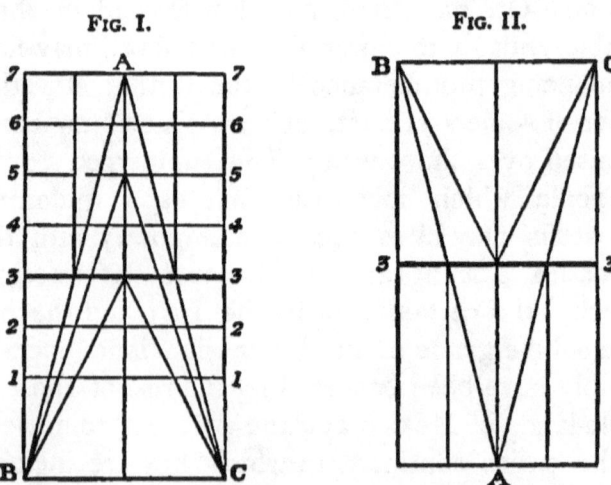

FIG. I. FIG. II.

Let the line 3, 3, in the diagram, Fig. I., represent that

of the sea's mean level; the line 3 C, or 3 B, the sea's mean depth; the triangle B A C, a rising continent; and the internal triangles, whose apices reach the lines 3, 3 and 5, 5 respectively, its comparative bulk or volume during its various intermediate stages of elevation. When the rising triangle (*i.e.*, continent) reaches the line 3, 3 (that of the sea-line ere the land began to rise), its mass, equal to that of the parallelogrammic band 1 B C 1, shall have displaced water to that amount, and sent it to the surface, which shall have risen, in consequence, from the line 3, 3 to the line 5, 5. When the continent reaches the line 5, 5, there will be another band, equal to half the mass of the first, displaced and sent to the surface, which shall now have risen to the line 6, 6; and not until the point of the triangle (*i.e.*, continent) has reached the line 7, 7, will it have overtaken the rising surface. Such, in proportion to its bulk, would be the effect, on the ocean-level of a rising continent, were there to be no equivalent sinking of the surface elsewhere,—just as, when the mercury of the thermometer is rising in the tube, there is no corresponding sinking of the metal contained in the instrument elsewhere, or, even if there *were* an equivalent sinking, were that sinking to take place in the interior of some immense tract of table-land.

Let us now, however, turn to the diagram Fig. II., and consider whether the full realization of the fiction of sinking hollows *within the sea*, exactly correspondent in their cubic contents to the rising continents, would be at all adequate to preserve the hypothetical fixity of ocean surface. Let the line B, C, Fig. II. represent the bottom of the ocean, and the triangle B, A, C, a depression of the earth's crust, exactly equal in cubic amount to the rising land in Fig. I., and taking place exactly at the same time. It will be at once seen, in running over the details, that even the hypothesis of balancing hollows formed in the sea as a set-off against the elevations, is wholly insufficient to establish the theory of a fixed line of sea-level. The hollow might be formed, and yet the level affected notwithstanding. Until the elevation had risen above the line 3, 3 in the diagram Fig. I., and the corresponding hollow sunk to the line 3, 3 in the diagram Fig. II., the surface-line would remain unaffected,—the water displaced by the rising eminence would be contained in the sinking hollow; but immediately as the land rose over the surface, there would be a portion of it—the sub-aërial portion—which would displace no water. The hollow, if it took place in the exact ratio of the elevation,—and such is the stipulated condition of the theory,—would receive after this point exactly double the quantity of water that the land displaced, and the line of the sea-level would fall. When the elevation would have risen to the point A of the one diagram, and the hollowing depression sunk to the point A of the other, the amount of water received over water displaced would be equal in quantity to one of the parallelogrammic bands, 1 2, 2 1, or 2 3, 3 2, Fig. I.; and the sea-level would in consequence sink to the line 2, 2. The exactly *balanced* hollow would fail to preserve the *balance*.

And so I cannot continue to hold as a first principle, that the line of the sea-level is a fixed and stable line; seeing that ere I could do so I should have to believe, *first*, that the earth's radius has undergone no diminution since the earliest geologic periods in which an ocean existed; *second*, that for every elevation which takes place on the surface of the globe there takes place a corresponding depression upon it elsewhere; *third*, that if the elevation takes place within the bed of the sea, the depression *also* takes place within the bed of the sea; and, *fourth*, that the elevations and depressions bear always a nicely-adjusted proportion to each other in their contents,—different at two different stages of their formation,—being up to a certain point exactly as one to one; and after that point has been reached, exactly as one to two. And I can find no adequate grounds for believing all this. But though it be thus far from self-evident that the mean level of the ocean is a fixed line, its rises and falls must have been slight indeed compared with those of the land. There are some of the Alps more than fifteen thousand feet in height; but, if spread equally over Europe, they would raise the general surface, says Humboldt, little more than twenty-one feet. And the displaced masses of the ocean, whether occasioned by the rising or the sinking of continents, have always to be spread over a surface thrice greater than that of *all* the land. A displacement, however, effected by the sinking of a continent which bore as large a proportion to the ocean as that borne by the Alps to Europe, would lower the general sea-line from the mean level of by far the best-marked of our ancient coast lines to the mean level of the existing one.

THE CHAIN OF CAUSES.

'IT is no recent discovery,' says an ingenious French writer of the last century, 'that there is no effect without a cause, and that often the smallest causes produce the greatest effects. Examine the situations of every people upon earth;—they are founded on a train of occurrences seemingly without connexion, but all connected. In this immense machine all is wheel, pulley, cord, or spring. It is the same in physical nature. A wind blowing from the southern seas and the remotest parts of Africa brings with it a portion of the African atmosphere, which, falling in showers in the valleys of the Alps, fertilizes our lands. On the other hand, our north wind carries our vapours among the negroes: we do good to Guinea, and Guinea to us. The chain extends from one end of the universe to the other.' Waiving, however, for the present, the moral view of the question, I may be permitted to present my readers with an illustration of the physical one,—*i.e.*, the dependence of the conditions of one country on the conditions on which some other and mayhap very distant country exists,—which may be new to some of them, and which the Frenchman just quoted could have little anticipated.

When in the island of Bute, to which I had gone on two several occasions in the course of a few weeks, in order to examine what are known to geologists as the Pleistocene deposits of the Kyles, my attention was directed to a deep excavation which had just been opened for the construction of a gas tank in the middle of the town of Rothesay. It was rather more than twenty feet in depth, and passed through five different layers of soil. First, passing downwards, there occurred about eighteen inches of vegetable mould, and then about seven feet of a partially consolidated ferruginous gravel, which rested on about eighteen inches more of peat moss,—once evidently a surface soil, like the overlying one, though of a different character,—abounding in what seemed to be the fragments of a rank underwood, and containing many hazel-nuts. Beneath this second soil there lay fully nine feet of finely stratified sea-sand; and under all, a bed of arenaceous clay, which the workmen penetrated to the depth of about two feet, but, as they had attained to the required depth of their excavation, did not pass through. And this bed of clay, at the depth of fully twenty feet from the surface, abounded in sea-shells,—not existing in the petrified condition, but, save that they had become somewhat porous and absorbent, in their original state. Not a few of them retained the thin brown epidermis, unchanged in colour; and the gaping and boring shells, whose nature it is to burrow in clay and sand, and which were present among them in two well-marked species, occupied, as shown by their position, the place in which they had lived and died. Now, of these ancient deep-lying shells, though a certain portion of them could be recognised as still British, there were proportionally not a few that no longer live within the British area;—in vain might the conchologist cast dredge for them in any sea that girdles the three kingdoms; and the whole, regarded as a group, differed from any other that exists in Europe in the present day. Ere, however, I pass on to decipher the record which they form, or translate into words the strange old prehistoric facts with which they are charged, let me briefly refer to the overlying deposits, and the successive periods of time which *they* seem to represent.

x

The upper layer of vegetable mould here fully exhausts the historic period. And yet the fine old town of Rothesay is not without its history. The ancient ivy-clad castle of the place is situated scarce a minute's walk from the excavation; the same stratum of vegetable mould lies around that forms the upper layer in the pit, furnishing rich footing to shrub and tree; and its green moat, deserted long since by the waters, was excavated of old in the ferruginous gravel. And yet, though compared with the age of the gravel-bed on which it stands, the date of its erection is as of yesterday : history fails to trace its origin : we only know that it was already an important stronghold in the days of Haco of Norway, one of whose captains besieged and took it,—that Robert III. of Scotland died broken-hearted within its walls,—and that it still furnishes with his second title the heir-apparent of the British throne. On many other parts of the coast, though apparently not here, this gravel-bed contains shells, all of which, unlike those of the arenaceous clay beneath, still live around our shores, and most of which occurred, ere the last upheaval of the land, as dead shells on the beaches of the old coast line. The old line itself, against which the sea seems to have stood for ages ere the final upheaval, is present here immediately behind the town, in an eminently characteristic form. Its precipices of rough conglomerate still exhibit the hollow lines, worn of old by the surf, and occupy such places in relation to the buildings below as prove that even the oldest erections of the town, with the first beginnings of the castle, were all raised on one of its wave-deserted beaches. But the annals of Rothesay, notwithstanding their respectable antiquity, or even such memorials of human origin in the neighbourhood as altogether extend beyond the memory of history, advance comparatively but a little way towards the period of the old coast line and the last upheaval. When, in the times of Julius Cæsar, Diodorus Siculus wrote his big gossiping history, St. Michael's Mount, in Cornwall, was connected with the mainland at low water as it is now,—a fact good in evidence to show that since that age the respective levels of land and water have not altered in Britain. The old coast line must have been already upheaved when Cæsar landed in the island. And yet, though, as shown by its profound caves and deeply excavated hollows, the sea must have beaten against it during an immensely protracted period of depression, there existed a previous period of upheaval, represented by the layer of moss at the bottom of the gravel, when the land must have stood considerably higher over the sea-level than it does now. In many localities around the shores of Britain and Ireland, the moss-bed which so often underlies the bed of old coast gravel is found to run out under the sea to depths never laid bare by the tide; and yet at least as low as the sea ever falls, it is found bearing its stumps and roots of bushes and trees of existing species, that evidently occupy the place in which they had originally grown and decayed. These submerged mosses, as they are termed, occur along the sides of the Firths of Tay and Forth, and in at least one locality on the southern side of the Moray Firth; on the west coast they lie deep in lochs and bays; they occur on various parts of the coasts of Ireland; and off the shores of Erris and Tyrawly have furnished a basis for strange legends regarding an enchanted land, which once in every seven years raises its head above the water, green with forests and fields, but on which scarce any one has succeeded in landing. They occur also on the English shores, in one interesting instance in the immediate neighbourhood of that St. Michael's Mount, which, from the description of the Sicilian historian, furnishes a sort of negative measure of the period during which the gravel bed immediately over them was elevated. 'On the strand of Mount's Bay, midway between the piers of St. Michael's Mount and Penzance, on the 10th of January 1757,' says

Borlase, in his *Natural History of Cornwall*, 'the remains of a wood, which anciently must have covered a large tract of ground, appeared. The sands had been drawn off from the shore by a violent sea, and had left several places, twenty yards long and ten wide, washed bare, strewed with stones like a broken causeway, and wrought into hollows somewhat below the rest of the sands. This gave me an opportunity of examining the following parts of the ancient trees :—In the first pool part of the trunk appeared, and the whole course of the roots, eighteen feet long and twelve wide, were displayed in a horizontal position. The trunk at the fracture was ragged ; and beside the level range of the roots which lay round it was part of the body of the tree, just above where the roots divided. Of what kind it was there did not remain enough positively to determine. The roots were pierced plentifully by the teredo or auger worm. Thirty feet to the west we found the remains of another tree : the ramifications extended ten feet by six ; there was no stock in the middle ; it was therefore part of the under or bottom roots of the tree, pierced also by the teredo, and of the same texture as the first. Fifty feet to the north of the first tree we found part of a large oak ; it was the body of a tree three feet in diameter; its top inclined to the east. We traced the body of this tree, as it lay shelving, the length of seven feet; but to what further depth the body reached we could not discern, because of the immediate influx of water as soon as we had made a pit for discovery. It was firmly rooted in earth six inches from the surface of the sand : not so fixed was the stock of a willow tree, with the bark on, one foot and a half in diameter, within two paces of the oak. The timber was changed into a ruddy colour; and hard by we found part of a hazel-branch, with its glossy bark on. The earth in all the tried places appeared to be a black, cold marsh, filled with fragments of leaves of the *Juncus aquaticus maxi-*

mus. The place where I found the trees was three hundred yards below full sea-mark. The water is twelve feet deep upon them when the tide is in.' It will be seen from this description,—and it agrees with that of our submerged forests of the period generally,—that the trees which grew on this nether soil, when the level of the land stood considerably higher than it does now, were exactly those of our present climate,—a fact borne specially out by the numerous hazel-nuts which the deposit almost everywhere contains. The hazel is one of the more delicate indigenous trees of the country. It was long ago remarked in Scotland by intelligent farmers of the old school, that 'a good *nut* year was always a good *oat* year;' and that 'as the *nut* filled the *oat* filled.' And now our philosophical botanists confirm the truthfulness of the observation embodied in these proverbial sayings, by selecting the hazel as the indigenous plant which most nearly resembles in its constitution the hardier cereals. It rises on our hill-sides to the height, but no higher, to which cultivation extends; and where the hazel would fail to grow, checked by the severity of the climate, it would be in vain to attempt rearing the oat, or to expect any very considerable return from either rye or barley. The existence of hazel nuts, then, in this mossy stratum, is fraught with exactly the same sort of evidence regarding the climate of that period of upheaval which it represents, as that borne by the shells of the overlying gravel to the subsequent period when the sea stood against the old coast line. Equally during both periods our country possessed its present comparatively genial climate,—the finest enjoyed by any country in the world situated under the same latitudinal lines. But the bed beneath gives evidence of an entirely different state of things.

Under the stratum of moss, as we have already said, there occurs in the Rothesay pit a thick bed of stratified sea-sand, and under the sand a bed of clay charged with shells; and these shells exist no longer as a group in the British seas. Regarded as characters charged with the climatal history of the period that represents the stratum in which they occur, the following list, with the attached explanations, may be regarded as indicative of the meanings which they bear. We may mention, that the greater number of the specimens specified were collected in the pit after our first visit to it, by Mr. John Richmond of the Temperance Hotel, Rothesay, to whose intelligent guidance and direction the geologic traveller, desirous of cultivating an adequate acquaintance with the Pleistocene deposits of the island in the least possible time, would do well to commit himself.

Natica clausa,	Not now a British species, but found living in the North Sea as far as Spitzbergen, and on the shores of boreal America.
Trophon scalariforme,	Not now British, but living in the same boreal seas as the other.
Buccinum Humphreysianum,	One of the rarest of British shells. 'It appears,' say Messrs. Forbes and Hanley in their History of the British Mollusca, 'to be an arctic form lingering in our fauna.'
Trochus inflatus,	Not now British: existing habitat unascertained.
Undescribed *natica,*	Ditto; ditto.
Trophon clathratus,	British; but also boreal.
Littorina rudis,	Ditto; ditto.
Tellina proxima,	Not yet found living in the British area; but abundant on the coasts of Greenland, boreal America, etc.
Saxicava sulcata,	Not now British.
Mya Uddevallensis,	Now deemed a variety of *Mya truncata,* but, save that it was found in one instance by Dr. Fleming among the Shetland Islands, not a British, but a boreal variety.

Undescribed *modiola*,	Not now British: existing habitat unascertained.
Mya truncata, *Saxicava rugosa*, *Lucina flexuosa*, *Astarte compressa*, *Nucula nucleus*,	British; but also boreal.

Such were the shells found in the arenaceous clay-bed of the Rothesay pit, full twenty feet from the surface; and from where, in various other parts of the country, the same bed has been reached by excavations, or found cropping out along the shores, the list has been greatly increased. At Balnakaillie Bay, for instance, in the Kyles of Bute, where Mr. Smith of Jordanhill,—one of our highest authorities on the Pleistocene formation,—first detected the deposit, we found several specimens of the *Pecten Islandicus*,—a fine shell, which, though abundant on the coast of Labrador, has not been found living on those of Britain; with specimens of *Panopea Norwegica*,—a massive shell, of the same boreal character, recently, however, found on our coast; though such is its extreme rarity, that a conchological friend tells us he was lately offered a British specimen for sale, at the not very moderate price of two pounds ten shillings. Even in the instance in which the shells are not only British, but also not of extreme rarity, the *proportions* in which they occur in the beds are certainly exotic. *Astarte elliptica*, for instance, is by no means a common Astarte in the Scottish seas, nor is it at all known in those of England or Ireland; whereas in Greenland it is very abundant; and in those beds in which it is the prevailing Astarte, it is in the Greenlandic, not in the Scottish proportions, in which it occurs. In the same way *Cyprina Islandica*, though comparatively rare in the Firth of Clyde, is not rare in the Scottish seas generally; but it is in the seas of Iceland, as its name implies, that it attains to its fullest numerical development; and in the Pleistocene beds of the Clyde it is in the Icelandic, not the Scottish proportions, that we find it. The same remark applies to *Cardium Norwegica* and *Astarte compressa*, with not a few others; and still more strongly to another Astarte, not rare in the Pleistocene deposits of at least Banffshire and Caithness, but so exceedingly rare as Scottish in the present age of the world, that the late Professor Edward Forbes,—indefatigable dredger as he was,— had to borrow from a friend the Scottish specimen which he figures in his great work. But though of such unfrequent occurrence in the Scottish seas, it is common in those of Nova Zembla and within the Arctic circle; and it is in the proportions in which it is developed in the high latitudes that we now find it in the Pleistocene beds of Scotland.

But how interpret so curious a fact as the occurrence in this country of beds of shells (evidently occupying the place in which they had lived and died) whose proper climatal habitat is now some ten or fifteen degrees further to the north? There is nothing more fixed than the nature of species. Art, within certain limits, exerts an acclimatizing power: Alpine plants may be found, for instance, living, if not flourishing, within the Botanic Gardens of Edinburgh, elevated scarce a hundred feet above the level of the sea; but every scientific gardener knows how extremely difficult it is to keep these alive in the too genial temperature of a situation greatly lower than the one natural to them; and that while inter-tropical plants may be easily maintained in existence through the judicious application of artificial heat, the sub-arctic or Alpine plants are ever and anon dying out. And never do they so change their natures as *of themselves* to propagate their kind downwards from the hill-tops to the plains. They on no occasion violate the climatal conditions imposed upon them by nature. It is so also with the animal world, and especially with shells. There are shells reckoned British, so delicately sensible of cold, that their northern limit barely touches the southern shores of Britain. That fine bivalve *Cytherea chione* is one of these, never getting further north than Caernarvon Bay; *Cardium rusticum*, so graphically described by Mr. Kingsley in his *Glaucus*, under the style and title of Signor Tuberculato, is another, ranging southward to the Canaries, but barely impinging, in its northern limits, on the shores of Devon and Cornwall; and our splendid Haliotus, or ear-shell, *H. tuberculata*, though reckoned British by courtesy, does not even touch the British shores, but finds its northern limit at the Channel Islands. Nor are the northern shells more tolerant of warm than the southern ones of cold water. We have already referred to *Astarte elliptica* as finding its southern line of boundary on the Scottish coasts; *Pecten Niveus* has not occurred to the south of the Firth of Clyde; and *Trochus undulatus*, though it ranges to Greenland, barely reaches our northern and western shores. Such and so nice is the dependence of shells on conditions of temperature, and such and so nice is their restriction to climatal areas. Nor could they have had a different nature in the past. How, then, could the cold *Natica clausa* and *Trophon scalariforme* of Spitzbergen and boreal America, and the *Tellina proxima* and *Mya Uddevallensis* of Greenland and the North Cape, have been at one time living denizens of the bay of Rothesay? Under what strange circumstances could whole scalps of the *Pecten Islandicus* have thriven in the Kyles of Bute, accompanied by groups of boreal *Saxicava*, that dug themselves houses in the stiff clay, and massive Panopea, that burrowed in the mud? The island of Bute is famous for now possessing perhaps the finest climate in Scotland: exotics blow in its gardens and shrubberies, that demand elsewhere the shelter of a green-house: and yet there was a time when, judging from the extreme boreal character of its shells, it pined under a severe and ungenial climate, in which even the hardier cereals could not have ripened. How account for a state of things so very unlike the present?

Questions in natural science cannot be resolved with all the certainty of questions in astronomical or mathematical science. Adams and Le Verrier could not only infer from the disturbances of Uranus the existence of a hitherto unknown planet, but even indicate its place in the heavens. But though the varying climatal circumstances of our country, and of northern Europe generally, seem to have depended scarce less surely on the varying physical conditions of another country three thousand miles away, than the irregularities of the planet Uranus did upon the mass and position of the planet Neptune, we question whether any amount of skill, or intimacy of acquaintance with the phenomena, could have led to an *a priori* anticipation of the fact. We shall afterwards show, however, that the climate of northern Europe is mainly dependent on the conditions of Northern America; and that one certain change in its condition gave to our country the severe climate which obtained when *Natica clausa* and *Tellina proxima* lived in the bay of Rothesay; and that it is a result of another certain change in its condition, that the delicate fuschia now expands its purple bells in Bute on the soil by which great deep-lying accumulations of these subarctic shells are covered.

Let us first remark, that during the period of the boreal shells the land was greatly depressed. The subsequent depression,—that represented in the Rothesay excavation by the upper gravel-bed,—that which succeeded the age of the submerged mosses,—that during which the waves broke against the old coast line,—seems to have been restricted to a descent of some thirty, or at most forty, feet beneath the level which the land at present maintains; whereas the previous depression,—that represented by the bed of arenaceous clay and the boreal shells,—must have been a depression of many hundred feet. No such inference, however, could be based on any of the Bute deposits which we have yet seen; and yet we might safely conclude, even from them, that when these deep-sea shells lived where we now find them, the land must have sat comparatively low in the water. When scalps of *Pecten Islandicus* throve on the argillaceous bed cut open above tide-mark by the little stream which falls into Balnakaillie Bay, and noble Panopea burrowed in its stiff clay, Bute must have existed, not as one, but as three islands, separated from each other by ocean sounds occupying the three valleys by which it is still traversed from side to side. In the neighbouring mainland many a promontory and peninsula must have also existed as detached islands. The long promontory of Cantyre and Knapdale, traversed by open sounds at Tarbert and Crinan, must have formed two of these; the larger part of the shire of Dumbarton, cut off from the mainland by straits passing inwards through the valleys of the Leven and of Loch Long, must also have borne an insular character; Loch Lomond must have existed, not as a fresh-water lake, but as an interior sea; and, in fine, the whole geography of the British islands must have been widely different from what it is now. There are other localities, however, in which, from the elevation of the boreal shell-bed over the present sea-level, we are justified in inferring that the depression of the land must have been much greater than that indicated by the beds of Bute. The same bed, and containing the same shells, was laid open in forming the Glasgow and Greenock Railway, a little to the west of Port-Glasgow, at an elevation of about fifty feet over the high-water line. It was detected at Airdrie, about fifteen miles inland, in the first instance, at a height of three hundred and fifty feet over the sea, and subsequently at the still more considerable height of five hundred and twenty-four feet. We ourselves have disinterred the same shells from where they rested, evidently *in situ*, in Banffshire,—on the top, in one instance, of a giddy cliff, elevated two hundred and thirty feet over the beach,—in another, lying deep in the side of a valley once a long withdrawing firth, but now fully six miles from the sea, and raised about a hundred and fifty feet above it. In Caithness they have been detected by Mr. Robert Dick at the greatest heights to which the boulder-clay attains; they occur also at very considerable heights in the boulder-clay of the Isle of Man; and were found by Mr. Trimmer in the drift of Moel Tryfon, in North Wales, at the extraordinary elevation over the sea of *fifteen hundred feet.* When the boreal shells at Airdrie lived, Scotland must have existed as a wintry archipelago, separated into three groups by the oceanic sounds of the great Caledonian Valley, and of the low flat valley, now traversed by the Union Canal, which extends between the Firths of Forth and Clyde. And when the shells of Moel Tryfon lived, only the higher parts of the Highlands of Scotland, and of the Cheviot and Lammermuir groups, could have had their heads elevated over the wintry ice-laden sea of the Pleistocene agents. There are grounds for holding that the period, though one geologically, was of vast extent,—that the degree of submergence was greater at one time and less at another; or, more strictly speaking, that the commencement of the period was one of gradual depression in the British area,—that about its middle term all was submerged, save the hill-tops and higher table-lands,—and that our country then began gradually to rise, until, about the close of the wintry *eon*, its level was mayhap scarce a hundred feet lower than it is at present. But though comparatively greater and less at different times, there seems to have been no time during the period, in which the depression was not absolutely great.

Let us next remark, as very important to our argument, that not only was the period one of depression in the British area, but also very extensively in the northern hemisphere generally. The shell-beds of Uddevalla,—identical in the character and species of their organisms with those of the Clyde,—are elevated two hundred feet above the neighbouring Cattegat; and in Russia Sir Roderick Murchison detected similar beds in the valley of the Dwina, lying nearly two hundred miles south-east of Archangel, and at least a hundred and fifty feet over the level of the White Sea. It is not uninteresting to mark, in the list of shells given by Sir Roderick in his great work on Russia, and which were the product, he states, of not more than two hours' exploration among these far inland beds, exactly the names of the same species that occurred in the Rothesay excavation, or may be found in the Pleistocene deposits of the Kyles. We recognise as the prevailing forms, *Natica clausa*, *Pecten Islandicus*, *Astarte elliptica*, *Astarte compressa*, *Mya truncata* in both its boreal and more ordinary varieties, and *Tellina proxima*, with many others. The inscriptions borne by the Pleistocene of both Sweden and Russia are formed of the same character as those exhibited by the Pleistocene of our own country, and tell exactly the same story. But it is of still more importance to our argument, that the Pleistocene of America is also inscribed with similar characters, and is coupled with similar evidence. Shell-beds identical in their contents with those of the Clyde, Uddevalla, and the valley of the Dwina, have been detected in the neighbourhood of Quebec, at the height of two hundred feet over the Atlantic, and traced onwards by Mr. Logan, the accomplished State-geologist for the Canadas, to the height of four hundred and sixty feet. And in these American beds, separated from those of the Dwina by a hundred and twenty degrees of longitude, *Pecten Islandicus*, *Natica clausa*, *Mya truncata*, *Saxicava rugosa*, and *Tellina proxima*, are the prevailing forms. How very wide the geographic area which these shells must have possessed of old! A depression of the North American Continent to the amount of but four hundred and sixty feet would greatly affect its contour. It would cut it off from Southern America—the highest point over which the Panama Railway passed was but two hundred and fifty feet over the level of the sea—and unite the Atlantic and Pacific Oceans by a broad channel, more than thirty fathoms in depth. But from various other appearances the American geologists claim for their country a much greater depression than even that of Moel Tryfon in Wales. It must have been depressed at least two thousand feet, and a wide sea must have passed through the valley of the Mississippi into what is now the Lake district, and from thence into Hudson's Bay and the Arctic Seas. And now, let the reader mark the probable effects on the climate of Northern Europe generally, and on that of Britain in particular, of so extensive a submergence of the American Continent.

No other countries in the world situated under the same lines of latitude enjoy so genial a climate as that enjoyed by the British islands in the present day. The bleak coasts of Labrador lie in the same parallels as those of Britain and Ireland; St. John's, in Newfoundland, is situated considerably to the south of Torquay in Devon; and Cape Farewell, in Greenland, to the south of Lerwick, the capital of the Shetland Islands. But how very different the climate of these bleak occidental lands, from that which renders Great Britain one of the first of agricultural countries! At Nain, in Labrador, situated in the same latitude as Edinburgh, the ground-frost at the depth of a few feet from the surface never thaws, but forms an ungenial rock-like subsoil, against which the labourer breaks his tool, and over which the cereals fail to ripen. From the northern coasts of Newfoundland, though lying under the same latitudinal lines as the extreme south of England, there forms in winter a thick cake of ice, which, binding up the stormy sea, runs northwards and eastwards, and connects, as with a long bridge, the north of Iceland with the north of Newfoundland; thus constituting a natural isothermal line, which shows that the European island has a not severer climate than the American one, though it lies more than ten degrees further to the north. And be it remembered that, did climate depend exclusively on a country's latitudinal position on the map, and its distance from the sun, it is the climate of Northern America that would be deemed the ordinary and proper climate, and that of Northern Europe the extraordinary and exceptional one. Great Britain and Ireland owe the genial, equable warmth that ripens year after year their luxuriant crops, and renders their winters so mild that the sea never freezes around their shores, *not*, at least directly, to the distant sun. Like apartments heated by pipes of steam or hot water, or green-houses heated by flues, they derive their warmth from a heating agent laterally applied: they are heated by warm water. The great Gulf Stream, which, issuing from the Straits of Florida, strikes diagonally across the Atlantic, and, impinging on our coasts, casts upon them not unfrequently the productions of the West Indies, and always a considerable portion of the warmth of the West Indies, is generally recognised as the heating agent which gives to our country a climate so much more mild and genial than that of any other country whatever similarly situated. Wherever its influence is felt,—and it extends as far north as the southern shores of Iceland, Nova Zembla, and the North Cape,—the sea in winter tells of its meliorating effects by never freezing: it remains open, like those portions of a reservoir or canal into which the heated water of a steam-boiler is supposed to escape. In some seasons,—an effect of unknown causes,—the Gulf Stream impinges more strongly against our coasts than at others: it did so in 1775, when Benjamin Franklin made his recorded observations upon it,—the first of any value which we possess; and again during the three mild winters that immediately preceded the last severe one,—that of 1855,—and which owed their mildness apparently to that very circumstance. It was found during the latter seasons that the temperature of the sea around our western coasts rose from one and a half to two degrees above its ordinary average; and it must be remembered how, during these seasons, every partial frost that set in at once yielded to a thaw whenever a puff of wind from the west carried into the atmosphere the caloric of the water over which it swept. The amount of heat discharged into the Atlantic by this great ocean-current is enormous. 'A simple calculation,' says Lieutenant Maury, 'will show that the quantity of heat discharged over the Atlantic from the waters of the Gulf Stream in a winter day would be sufficient to raise the whole column of atmosphere that rests upon France and the British Islands from the freezing point to summer heat. It is the influence of this stream upon climate,' he adds, 'that makes Erin the Emerald Isle of the sea, and clothes the shore of Albion with evergreen robes; while, in the same latitude, on the other side, the shores of Labrador are fast bound in fetters of ice.'

Now, a depression beneath the sea of the North American continent would have the effect of depriving northern Europe of the benefits of this great heating current. Its origin has been traced to various causes,—some of them very inadequate ones. It has been said, for instance, that it is but a sort of oceanic prolongation of the Mississippi. It has been demonstrated, however, that it discharges through the Straits of Florida about a thousand times more water than the Mississippi does at its mouth; and yet, even were the case otherwise, and the view correct, any great depression of North America would cut off the Mississippi from among the list of great rivers, by converting the valley which it occupies into a sea, and would thus terminate the existence of the Gulf Stream. The stream has, however, a very different and more adequate origin, but one which the depression of the North American continent would equally affect. It is a reaction on the great Drift Current. If the reader take a cup or basin filled with water, and blow strongly across the surface of the fluid, two distinct currents will be generated,—a *drift* current, which, flowing in the direction of his breath, will impinge against the opposite side of the vessel,—and a reactionary current, which, passing along its sides, will return towards himself. And nothing can be more obvious than the principle on which this occurs. The *drift* current, more immediately generated by his breath, heaps up the water against the side of the vessel on which it impinges; and this heaped-up water must of course inevitably seek to return to the other side, in order to restore the deranged equilibrium of level. Now, the Northern Atlantic,—the Atlantic to the north of the equator,—displays on an immense scale exactly the phenomena exhibited by this simple experiment of the cup or basin. The breath of the trade-winds, ever blowing upon it from the east and north-east, in that broad belt which lies between the tenth and the twenty-sixth degrees of north latitude, forms a great drift current, which, impinging on and heaping up the waters against the South American coast,—the opposite side of the cup or basin,—flows northwards into the Carribbean Sea and Mexican Gulf, and, issuing from the Straits of Florida in the character of the reactionary Gulf Stream, strikes diagonally across the Atlantic full on Northern Europe. But the existence of this reactionary stream is not merely and exclusively a consequence of the existence of the Drift Current: it is also equally a consequence of the existence of an American continent. Save for the side of the basin or cup opposite to that whence the breath comes, the water, instead of returning in a reactionary current, would flow over. Such a wide breach in the sides of the *cup* along the Isthmus of Panama, for instance, as a depression of but four hundred and sixty feet would secure, would permit the Drift Current to flow into the Pacific. Such a wide breach in the sides of the *cup* along the Valley of the Mississippi as a depression equal to that indicated by the shells of Moel Tryfon would secure, would permit the reactionary Gulf Stream, though already formed, to escape, along what is now the lake district of America, into Hudson's Bay. In either case the Gulf Stream would be lost to Northern Europe; and the British Islands, robbed of the Gulf Stream, would possess merely the climate proper to their latitudinal position on the map; —they would possess such a climate as that of Labrador, where, beneath seas frozen over every winter many miles from the shore, exactly the same shells now live as may be found, in the sub-fossil state, in the Kyles of Bute, or underlying the pleasant town of Rothesay. A submergence of the North American continent would give to Britain and Ireland, with the countries of Northern Europe generally, what they all seem to have possessed during the protracted ages of the Pleistocene era,—a glacial climate.

If our conclusions be just,—and we see not on what grounds they are to be avoided,—our readers will, we daresay, agree with us that it would not be easy to produce a more striking illustration of the influences which are at times exerted by the conditions of one country on those of another. Our brethren of the United States are occasionally not a little jealous of the mother country; but we suspect all of them do not know how completely they could ruin her could they but succeed in keeping their great Gulf Stream to themselves. It might be unwise, however, to urge matters quite so far, lest they should provoke us, in turn, to demand back again the large brains and high-mettled blood which we have most certainly given them. Such of our readers as occasionally enjoy a summer vacation on the west coast might find it no dull or useless employment to begin reading for themselves the shell inscriptions borne by the Pleistocene deposits. It would at once form an excellent exercise in Conchology and a first lesson in Geology, which, from the interest it excited, would scarce fail to lead on to others. With their eyes educated to the work too, they would find, we doubt not, the beds in many a new locality in which they had not been detected before; and enjoy the same sort of pleasure in falling upon a fresh deposit, as that enjoyed by an Egyptian or Assyrian antiquary when he discovers a catacomb of unrolled mummies never before laid open, or a series of sculptures or of inscriptions in the cuneiform character, unseen since the days of Semiramis or Sennacherib. We ourselves once enjoyed such a pleasure at Fairlie;—we laid open a noble bed, previously unknown, about a quarter of a mile to the north of the village; and from amid great scalps of *Pecten Islandicus*, roughened on their upper valves by huge Balonidæ, and from beside thick-lying groups of Cypinidæ, we disinterred many a curious boreal shell,—great massive Panopea, graceful Veneridea, the Greenland Mya, and the Tellina of the North Cape; and beneath all we detected grooved and dressed rock-surfaces, that bore their significant markings as freshly as if the grating ice had passed over them but yesterday. We would specially call the explorer's attention to the corroborative evidence borne by appearances of mechanical origin such as these to the mute testimony of the shells. We have already incidentally referred to the interesting deposits of Balnakaillie Bay. A stream falls into the sea at its upper extremity, and exhibits, in the section which it supplies, a bed charged with the old boreal shells, from where it creeps out along the beach, till where we lose it in the interior, far above the reach of the tide. As it passes inwards, we find the old coast line deposits resting over it; in one place assuming the ordinary character of a stratified sand and gravel; in another existing as a partially consolidated conglomerate; while immediately beneath it, on the north side of the stream, the rock appears strongly marked by the old glacial dressings. The mechanical and zoologic evidences of the existence of a period of extreme cold thus lying side by side may be studied together. But the district has its many such appearances. Not a few of the hills bear, in their rounded protuberances and smoothed and channelled hollows, evidence of the ice-agent that wasted them of old; and in the valley of the Gareloch, only a few miles distant, Mr. Charles Maclaren found unequivocal traces of an ancient glacier.

But the collateral evidences would lead us into a field quite as wide as that into which we have made our brief excursion, and are now preparing to leave. The following interesting extract from Mr. Kingsley's *Glaucus*, with which we conclude, may at once show how rightly to read these, and what very amusing reading they form. It is thus we find Mr. Kingsley accounting, in light and graceful dialogue, for the formation of a profoundly deep lochan of limited area, that opens its blue eye to the heavens amid the rough wilderness of rocks and hills that encircle the gigantic Snowdon.

'You see the lake is nearly circular: on the side where we stand the pebbly beach is not six feet above the water, and slopes away steeply into the valley behind us, while before us it shelves gradually into the lake; forty yards out, as you know, there is not ten feet water, and then a steep bank, the edge whereof we and the big trout know well, sinks suddenly to unknown depths. On the opposite side, that vast flat-topped wall of rock towers up shoreless into the sky seven hundred feet perpendicular: the deepest water of all, we know, is at its very foot. Right and left two shoulders of down slope into the lake. Now turn round, and look down the gorge. Remark that the pebble bank on which we stand reaches some fifty yards downward: you see the loose stones peeping out everywhere. We may fairly suppose that we stand on a dam of loose stones, a hundred feet deep.

'But why loose stones? and if so, what matter and what wonder? There are rocks cropping out everywhere down the hill-side.

'Because, if you will take up one of these stones, and crack it across, you will see that it is not of the same stuff as those said rocks. Step into the next field and see. That rock is the common Snowdon slate which we see everywhere. The two shoulders of down right and left are slate too; you can see that at a glance. But the stones of the pebble bank are a close-grained yellow-spotted Syenite; and where,—where on earth did these Syenite pebbles come from? Let us walk round to the cliff on the opposite side and see.

'Now mark. Between the cliff-foot and the sloping down is a crack, ending in a gully: the nearer side is of slate, and the further side the cliff itself. Why, the whole cliff is composed of the very same stone as the pebble ridge.

'Now, my good friend, how did these pebbles get three hundred yards across the lake? Hundreds of tons, some of them three feet long,—who carried them across? The old Cimbri were not likely to amuse themselves by making such a breakwater up here in No-man's-land, two thousand feet above the sea; but somebody or something must have carried them, for stones do not fly, nor swim neither.

'Let our hope of a solution be in John Jones, who carried up the coracle. Hail him, and ask what is on the top of that cliff. So?—"Plains and bogs, and another linn." Very good. Now, does it not strike you that the whole cliff has a remarkably smooth and plastered look, like a hare's run up an earth bank? And do you see that it is polished thus only over the lake? that as soon as the cliff abuts on the downs right and left, it forms pinnacles, caves, broken angular boulders? Syenite usually does so in our damp climate, from the weathering effect of frost and rain; why has it not done so over the lake? On that part something (giants perhaps) has been scrambling up and down on a very large scale, and so rubbed off every corner which was inclined to come away, till the solid core of the rock was bared. And may not these mysterious giants have had a hand in carrying the stones across the lake? . . . Really I am not altogether jesting. Think a while, what agent could possibly have produced either one or both of these effects?

'There is but one; and that, if you have been an Alpine traveller, much more if you have been a chamois-hunter, you have seen many a time (whether you knew it or not) at the very same work.

'Ice! Yes; ice. Hrymin the frost-giant, and no one else. And if you look at the facts, you will see how ice may have done it. Our friend John Jones' report of plains and bogs, and a lake above, makes it quite possible that in the ice-age (glacial epoch, as the big-word-mongers call it), there was above that cliff a great neve or snow-field, such as you have seen often in the Alps at the head of each glacier. Over the face of this cliff, a glacier had crawled down from that neve, polishing the face of the rock in its descent; but the snow, having no large and deep outlet, has not slid down in a sufficient stream to reach the vale below, and form a glacier of the first order, and has therefore stopped short on the other side of the lake, as a glacier of the second order, which ends in an ice-cliff hanging high up on the mountain side, and kept from further progress by daily melting. If you have ever gone up the Mer-de-Glace to the Tacul, you saw a magnificent specimen of the sort on your right hand, just opposite the Tacul, in the Glacier de Trélaporte, which comes down from the Aiguille de Charmoz.

'This explains our pebble ridge. The stones which the glacier rubbed off the cliff beneath it, it carried forwards slowly but surely, till they saw the light again in the face of the ice-cliff, and dropped out of it under the melting of the summer sun, to form a huge dam across the ravine; till, the "ice-age" past, a more genial climate succeeded, and neve and glacier melted away; but the "moraine" of stones did not, and remain to this day, the dam which keeps up the waters of the lake.

'There is my explanation. If you can find a better, do; but remember always that it must include an answer to,— How did the stones get across the lake?'

RECENT GEOLOGICAL DISCOVERIES.

Preparations for the British Association Meeting at Aberdeen in 1859.

THE gentlemen of the hammer and chisel must immediately prepare a *Reform Bill*, and re-adjust their nomenclature and classification. Both are uncouth and barbarous as well as unscientific. Recent discoveries have unsettled almost every one of the characters and tests of the age of rocks. Old Werner's Transition class, though founded to some extent on facts, has been long ago discarded. But will hardness or crystalline structure, or the absence even of organic remains, hitherto described as the grand features of the primitive class of rocks, now bear to be trusted as essentialia of classification? Every summer's ramble multiplies proofs to the contrary. The mere vicinity of a trap-vein, squirted from its boiling caldron below, among the most sedimentary strata, has often baked them into hard crystalline masses, and converted mud-banks charged with shells into beautiful granular marble, as may be seen at Strath, in Skye, under the overlying igneous rocks of the Cuchullins. And perhaps the time is not far distant when it may be difficult to find in the crust of the globe any assemblage of rocks in which organisms may not be detected, although *heat*, for the most part, has nearly obliterated them.[1] Again, a little more patient investigation, we expect, will

---

[1] 'The hypothesis,' says Sir Roderick Murchison, in his newly-published edition of *Siluria*, 'that all the earliest sediments have been so

blow to the winds many a fine theory as to the *gradual development* of species, and will most likely show that at no former period was there an ocean replete with shells and worms low in the scale of organization, which had not on its shores a rich vegetation and a *fauna* abounding in reptiles, and perhaps birds and quadupeds! Thus, when Hugh Miller wrote his *Old Red Sandstone*, he described it as peculiarly a *salt-water fish* formation, in which there was scarcely any shells or vegetables, the faint traces of the latter which he had discovered being only markings of fucoids and similar sea-weeds. So far as then known, the Scottish Old Red Sandstone was the produce of a deep shoreless ocean, to which no decayed forests had been brought down by rains and rivers to become future coal-fields, nor on whose margins and lagunes disported the amphibious crocodile or other allied genera, who could leave the impress of their feet or tails on the soft mud or sand. The formation, in short, was considered very low down indeed, and near the base of the platform of rocks in which rest entombed altered as to have obliterated the traces of any relics of former life which may have been entombed in them, is opposed by examples of enormously thick and often finely levigated deposits beneath the lowest fossiliferous rocks, and in which, if many animal remains had ever existed, more traces of them would be detected.'

'And yet,' as he again observes, 'the fine aggregation and unaltered condition of those sediments have permitted the minutest impressions to be preserved. Thus, not only are the broad wave-marks distinct, but also those smaller ripples which may have been produced by wind, together with apparent rain-prints, as seen upon the muddy surface, and even cracks produced by the action of the sun on a half-dried surface. Again, as a further indication that these are littoral markings, and not the results of deep-sea currents, the minute holes left by the Annelides are most conspicuous on the sheltered sides of the reptiles in each slab.

'Surely, then, if animals of a higher organization had existed in this very ancient period, we should find their relics in this sediment, so admirably adapted for their conservation, as seen in the markings of the little arenicola, accompanied even by the traces of diurnal atmospheric action.'—*Siluria*, pp. 20-27.

L. M.

the remains of the earliest races of organized creatures. But what have the discoveries of the last six months established? Why, this, that the Old Red Sandstone of the east coast of Scotland is comparatively a *modern* formation,—much newer, at least, than the grand and lofty masses of the purple and red conglomerate of the western coast, which they so greatly resemble, but upon which Sir Roderick Murchison has now proved that an extensive series of crystalline quartz-rocks, limestones, and micaceous schists repose, all *greatly older* than Hugh Miller's fish-beds! The discovery a few years ago of a little frog-like, air-breathing reptile in Morayshire (named the *Telerpeton Elginense*), has been a bone of contention among the *savans*, because, according to past theories, it was not easy to admit that it could have lived at the date of the deposition of the Old Red Sandstone; and hence very grave doubts were expressed about it, and much anxiety shown to establish that it belonged to the *carboniferous* strata, or to a *New Red* Sandstone formation, which, if it did exist in our district, would be most valuable, from the salt and calcareous deposits in which it usually abounds. But within the last month or so, Sir Roderick Murchison, in company with the Rev. G. Gordon of Birnie, made transverse sections of the whole series of Morayshire freestones, from the edges of the micaceous schist in the interior, to the maritime promontories of Burghead and Lossiemouth, which convinced them that the whole red and yellowish sandstones of the province 'are so bound together by mineral characters and fossil remains, that they must all be grouped as *Old Red* or *Devonian*.' Nay, more than this, the views of the Director-General of the Geological Survey have been confirmed and extended by the further discovery of *foot-prints* in the Burghead sandstone, not only of a small reptile like the Telerpeton, but of very large creatures, that in their movements made enormous strides, and whose bushy tails have left *trails* more distinct than the largest seals or otters could do! A well-known labourer in the English deposits (S. H. Beckles, Esq.), whose discoveries, in the Purbeck and Wealden beds, of the jaw-bones of most gigantic reptilia, have been extensive and most important, has recently examined the sandstone quarries at Burghead and Covesea, where he has discovered the most undoubted *foot-prints* of both large and small animals; and he has sent an extensive set of specimens to London, to be laid before the Geological Society at its winter meetings. Other *foot-marks* (each having the impression of three or four claws to it) have lately been seen by Sir Roderick, Mr. Martin of Elgin, and Mr. Gordon, and specimens communicated by Mr. P. Duff; so that, in the language of Sir Roderick Murchison's announcement to the late meeting of the British Association at Leeds, 'the presence of large reptiles, as well as of the little Telerpeton, in this upper member of the Old Red Sandstone, is completely established.'

We have not room enough at present to point out further deductions from these facts, and from the discovery, about three years ago, of *Silurian* fossils in the Southern Highlands and in Ayrshire. We allude to them only to show that, as in the days of Hutton and Playfair, the granite veins which traverse in all directions the schists of Glen-Tilt were the means of establishing the irruptive and igneous origin of granite, so Scotland again turns out to be the battle-field of our men of science, and that very great things may be expected from the explorations which undoubtedly will be made, in connexion with next meeting of the Association, to be held next autumn at Aberdeen, under the eye of the Prince Consort, and at which Sir Roderick Murchison, we are glad to understand, is to take his place as vice-president in all the sections. He is the senior of the three permanent trustees of the Association, and one of the founders of the body in 1831, of whom, strange to say, only five are now alive. In Sir David Brewster the science of the south of Scotland will be admirably represented and supported; while Sir Roderick, a Ross-shire man, an alumnus of the Inverness Academy (ay, and one who put shoulder to shoulder with the Highlanders on Corunna's bloody sod), will represent the land north of the Spey.

If we might suggest to those who will take the lead in the arrangements for the Aberdeen meeting, we would say that they ought, in the geological section, to prepare for one excursion to Stonehaven, on the eastern coast, and another to Cromarty and Eathie, the scenes of Hugh Miller's labours, on the north.

In Stonehaven bay, and arising out of the harbour, may be seen large dykes of trap ascending the cliff and overspreading the sandstone strata like the branches of a palm-tree, and thence overflowing towards the very curious quartzose conglomerate at Dunnottar Castle. On the other or northern horn of the bay, irruptive or felspar rocks jut up in great masses and promontories, shifting and disturbing the sandstone strata; and immediately beyond, these latter give place to hard crystalline and vertical strata, as to which the Association will have to decide whether they are altered *Silurian* or true *primitive* rocks.

At Cromarty, the local authorities, we think, should prepare for a visit from a large body of *savans* (which our railway and steamers will render easy), by exploring some new sections of the rocks on which Hugh Miller used to work. Many of these, it is well known, are below high-water mark, and are thus often covered by the sea; while almost all the nodules containing fossil-fish have been extracted and carried away. Some excavations in the *strike* or *line* of the same rocks should be made inland, the gravel and boulder-clay should be removed, a few layers of the sandstone underneath loosened, and a few broad sheets of the rock exposed *in situ*, and so left for the further examination of visitors,

EXTRACT FROM 'FOOTPRINTS OF THE CREATOR,' P. 199.

IN my little work on the Old Red Sandstone, I have referred to an apparent lignite of the Lower Old Red of Cromarty, which presented, when viewed by the microscope, marks of the internal fibre. The surface, when under the glass, resembled, I said, a bundle of horse-hairs lying stretched in parallel lines; and in this specimen alone, it was added, had I found aught in the Lower Old Red Sandstone approaching to proof of the existence of dry land. About four years ago, I had this lignite put stringently to the question by Mr. Sanderson; and deeply interesting was the result. I must first mention, however, that there cannot rest the shadow of a doubt regarding the place of the organism in the geologic scale. It is unequivocally a fossil of the Lower Old Red Sandstone. I found it partially embedded, with many other nodules half-disinterred by the sea, in an ichthyolitic deposit, a few hundred yards to the east of the town of Cromarty, which occurs more than four hundred feet over the Great Conglomerate base of the system. A nodule that lay immediately beside it contained a well-preserved specimen of the *Coccosteus decipiens;* and in the nodule in which the lignite itself is contained the practised eye may detect a scattered group of scales of *Diplacanthus,* a scarce less characteristic organism of the lower formation. And what, asks the reader, is the character of this ancient vegetable,—the most ancient, by three whole formations, that has presented its internal structure to the microscope? Is it as low in the scale of development as in the geological scale? Does this venerable Adam of the forest appear, like the Adam of the infidel, as a squalid, ill-formed savage, with a rugged shaggy nature, which it would require the suggestive necessities of many ages painfully to lick into civilisation? Or does it appear rather like the Adam of the poet and the theologian, independent, in its instantaneously-derived perfection, of all after development,—

> 'Adam, the goodliest man of men since born
> His sons?'

Is this tissue vascular or cellular, or, like that of some of the cryptogamia, intermediate? Or what, in fine, is the nature and bearing of its mute but emphatic testimony on that doctrine of progressive development[1] of late so strangely resuscitated?

In the first place, then, this ancient fossil is a true wood, —a dicotyledonous or polycotyledonous *Gymnosperm*, that, like the pines and larches of our existing forests, bore naked seeds, which, in their state of germination, developed either double lobes to shelter the embryo within, or shot out a fringe of verticillated spikes, which performed the same protective functions, and that, as it increased in bulk year after year, received its accessions of growth in outside layers. In the transverse section the cells bear the reticulated appearance which distinguish the coniferæ; the lignite had been exposed in its bed to a considerable degree of pressure; and so the openings somewhat resemble the meshes of a net that has been drawn a little awry; but no general obliteration of their original character has taken place, save in minute patches, where they have been injured by compression or the bituminizing process. All the tubes indicated by the openings are, as in recent coniferæ, of nearly the same size; and though, as in many of the more ancient lignites, there are no indications of annual rings, the direction of the medullary rays is distinctly traceable. The longitudinal sections are rather less distinct than the transverse one: in

[1] This alludes, of course, to the *development* theory of the *Vestiges of the Natural History of Creation.*

the section parallel to the radius of the stem or bole the circular disks of the coniferæ were at first not at all detected : and, as since shown by a very fine microscope, they appear simply as double and triple lines of undefined dots, that somewhat resemble the stippled markings of the miniature painter ; nor are the openings of the medullary rays frequent in the tangental section (*i.e.* that parallel to the bark) ; but nothing can be better defined than the peculiar arrangement of the woody fibre, and the longitudinal form of the cells. Such is the character of this the most ancient of lignites yet found that yields to the microscope the peculiarities of its original structure. We find in it an unfallen *Adam*,—not a half-developed savage.[1]

The olive-leaf which the dove brought to Noah established

---

[1] On a point of such importance I find it necessary to strengthen my testimony by auxiliary evidence. The following is the judgment, on this ancient petrifaction, of Mr. Nicol of Edinburgh,—confessedly one of our highest living authorities in that division of fossil botany which takes cognizance of the internal structure of lignites, and decides, from their anatomy, their race and family :—

'EDINBURGH, 19*th July* 1845.

' DEAR SIR,—I have examined the structure of the fossil-wood which you found in the Old Red Sandstone at Cromarty, and have no hesitation in stating, that the reticulated texture of the transverse sections, though somewhat compressed, clearly indicates a coniferous origin ; but as there is not the slightest trace of a disk to be seen in the longitudinal sections parallel to the medullary rays, it is impossible to say whether it belongs to the pine or araucarian division.— I am, etc.

' WILLIAM NICOL.'

It will be seen that Mr. Nicol failed to detect what I now deem the disks of this conifer,—those stippled markings to which I have referred. But even were this portion of the evidence wholly wanting, we would be left in doubt, in consequence, not whether the Old Red lignite formed part of a true gymnospermous tree, but whether that tree is now represented by the pines of Europe and America or by the araucarians of Chili and New Zealand. Were I to risk an opinion in a department not particularly my province, it would be in favour of an araucarian relationship.

at least three important facts, and indicated a few more. It showed most conclusively that there was dry land, that there were olive-trees, and that the climate of the surrounding region, whatever change it might have undergone, was still favourable to the development of vegetable life. And, further, it might be very safely inferred from it, that if olive-trees had survived, other trees and plants must have survived also; and that the dark muddy prominences round which the ebbing currents were fast sweeping to lower levels would soon present, as in antediluvian times, their coverings of cheerful green. The olive-leaf spoke not of merely a partial, but of a general vegetation. Now, the coniferous lignite of the Lower Old Red Sandstone we find charged, like the olive-leaf, with a various and singularly interesting evidence. It is something to know, that in the times of the *Coccosteus* and *Asterolepis* there existed dry land, and that that land wore, as at after periods, its soft, gay mantle of green. It is something also to know, that the verdant tint was not owing to a profuse development of mere immaturities of the vegetable kingdom,—crisp, slow-growing lichens, or watery spore-propagated fungi, that shoot up to their full size in a night,—nor even to an abundance of the more highly organized families of the liverworts and the mosses. These may have abounded then as now; though we have not a shadow of evidence that they did. But while we have no proof whatever of *their* existence, we have conclusive proof that there existed orders and families of a rank far above them. On the dry land of the Lower Old Red Sandstone, on which, according to the theory of Adolphe Brongniart, nothing higher than a lichen or a moss could have been expected, the ship-carpenter might have hopefully taken axe in hand to explore the woods for some such stately pine as the one described by Milton,—

'Hewn on Norwegian hills, to be the mast
Of some great ammiral.'

SIR RODERICK MURCHISON ON THE RECENT GEOLOGICAL DISCOVERIES IN MORAYSHIRE.

At a meeting of the Geological Society of London, held on the 15th December 1858, Part III. of a paper by Sir Roderick Murchison, on 'The Geological Structure of the North of Scotland,' was read.

Referring to his previous memoir for an account of the triple division of the Old Red Sandstone of Caithness and the Orkney Islands, Sir Roderick showed how the chief member of the group in those tracts diminished in its range southwards into Ross-shire, and how, when traceable through Inverness and Nairn, it was scarcely to be recognised in Morayshire, but re-appeared, with its characteristic ichthyolites, in Banffshire (Dipple, Tynet, and Gamrie). He then prefaced his description of the ascending order of the strata belonging to this group in Morayshire by a sketch of the successive labours of geologists in that district; pointing out how, in 1828, the sandstones and cornstones of this tract had been shown by Professor Sedgwick and himself to constitute, together with the inferior Red Sandstone and conglomerate, one natural geological assemblage; that in 1839 the late Dr. Malcolmson made the important additional discovery of fossil fishes, in conjunction with Lady Gordon Cumming; and also read a valuable memoir on the structure of the tract, before the Geological Society, of which, to his the author's regret, an abstract only had been published (*Proc. Geol. Soc.* vol. iii. p. 141). Sir Roderick revisited the district in the autumn of 1840, and made sections in the environs of Forres and Elgin. Subsequently, Mr. P. Duff of Elgin published a 'Sketch of the Geology of

Moray,' with illustrative plates of fossil-fishes, sections, and a geological map, by Mr. John Martin; and afterwards Mr. Alexander Robertson threw much light upon the structure of the district, particularly as regarded deposits younger than those under consideration. All these writers, as well as Sedgwick and himself, had grouped the yellow and whitish-yellow sandstones of Elgin with the Old Red Sandstone; but the discovery in them of the curious small reptile, the *Telerpeton Elginense*, described by Mantell in 1851 from a specimen in Mr. P. Duff's collection, first occasioned doubts to arise respecting the age of the deposit. Still, the sections by Captain Brickenden, who sent that reptile up to London, proved that it had been found in a sandstone which dipped under 'Cornstone,' and which passed downwards into the Old Red series. Captain Brickenden also sent to London natural impressions of the foot-prints of an apparently reptilian animal in a slab of similar sandstone, from the coast-ridge extending from Burghead to Lossiemouth (Cummingston). Although adhering to his original view respecting the age of the sandstones, Sir R. Murchison could not help having misgivings and doubts, in common with many geologists, on account of the high grade of reptile to which the Telerpeton belonged; and hence he revisited the tract, examining the critical points, in company with his friend the Rev. G. Gordon, to whose zealous labours he owned himself to be greatly indebted. In looking through the collections in the public Museum of Elgin, and of Mr. P. Duff, he was much struck with the appearance of several undescribed fossils, apparently belonging to reptiles, which, by the liberality of their possessors, were, at his request, sent up for inspection to the Museum of Practical Geology. He was also much astonished at the state of preservation of a large bone (*ischium*) apparently belonging to a reptile, found by Mr. Martin in the same sandstone quarries of Lossiemouth in which the scales or scutes of the Stagonolepis, described as belonging to a fish by Agassiz, had been found. On visiting these quarries, Mr. G. Gordon and himself fortunately discovered other bones of the same animal; and these, having been compared with the remains in the Elgin collections, have enabled Professor Huxley to decide that, with the exception of the Telerpeton, all these casts, scales, and bones belong to the reptile *Stagonolepis Robertsoni.* Sir Roderick, having visited the quarries in the coast-ridge, from which slabs with impressions of reptilian foot-marks had long been obtained, induced Mr. G. Gordon to transmit a variety of these, which are now in the Museum of Practical Geology, and of which some were exhibited at the meeting.

After reviewing the whole succession of strata, from the edge of the crystalline rocks in the interior to the bold cliffs on the sea-coast, the author has satisfied himself that the reptile-bearing sandstones must be considered to form the uppermost portion of the Old Red Sandstone, or Devonian group, the following being among the chief reasons for his adherence to this view :—1. That these sandstones have everywhere the same strike and dip as the inferior red sandstones containing Holoptychii and other Old Red ichthyolites, there being a perfect conformity between the two rocks, and a gradual passage from the one into the other. 2. That the yellow and light colours of the upper band are seen in natural sections to occur and alternate with red and green sandstones, marls, and conglomerates low down in the ichthyolitic series. 3. That whilst the concretionary limestones called 'Cornstones' are seen amidst some of the lowest red and green conglomerates, they re-appear in a younger and broader zone at Elgin, and re-occur above the Telerpeton-stone at Spynie Hill, and above the Stagonolepis-sandstone of Lossiemouth; thus binding the whole into one natural physical group. 4. That whilst the small patches of so-called 'Wealden' or Oolitic strata, described by Mr. Robertson and others, occurring in this district, are wholly unconformable to, and rest upon, the eroded surfaces of all the rocks under consideration, so it was shown that none of the Oolitic or Liassic rocks of the opposite side of the Moray Firth, or those of Brora, Dunrobin, Eathie, etc., which are charged with Oolitic and Liassic remains, resemble the reptiliferous sandstones and 'Cornstones' of Elgin, or their repetitions in the coast-ridge that extend from Burghead to Lossiemouth. *Fully aware of the great difficulty of determining the exact boundary-line between the Uppermost Devonian and Lowest Carboniferous strata, and knowing that they pass into each other in many countries, the author stated that no one could dogmatically assert that the reptile-bearing sandstones might not, by future researches, be proved to form the commencement of the younger era.*

Sir Roderick concluded by stating that the conversion of the Stagonolepis into a reptile of high organization, though of nondescript characters, DID NOT INTERFERE WITH HIS LONG-CHERISHED OPINION—FOUNDED ON ACKNOWLEDGED FACTS—AS TO THE PROGRESSIVE SUCCESSION OF GREAT CLASSES OF ANIMALS, *and that, inasmuch as the earliest trilobite of the invertebrate Lower Silurian era was as wonderfully organized as any living Crustacean, so it did not unsettle his belief to find that the earliest reptiles yet recognised,—the Stagonolepis and Telerpeton,—pertained to a high order of that class.*

At the same meeting, papers were read 'On the *Stagonolepis Robertsoni* of the Elgin Sandstones, and on the Footmarks in the Sandstones of Cummingston,' by Mr. T. H. Huxley; as well as one 'On Fossil Foot-prints in the Old Red Sandstones at Cummingston,' by S. H. Beckles, Esq.

www.ingramcontent.com/pod-product-compliance
Lightning Source LLC
Chambersburg PA
CBHW032018220426
43664CB00006B/289